DEFENSE, SECURITY AND STRATEGIES

PREVENTION OF NUCLEAR SMUGGLING AND TERRORIST TRAVEL

U.S.-FOREIGN PARTNERSHIP EFFORTS

DEFENSE, SECURITY AND STRATEGIES

Additional books in this series can be found on Nova's website under the Series tab.

Additional E-books in this series can be found on Nova's website under the E-book tab.

DEFENSE, SECURITY AND STRATEGIES

PREVENTION OF NUCLEAR SMUGGLING AND TERRORIST TRAVEL

U.S.-FOREIGN PARTNERSHIP EFFORTS

VALERE PRUITT
EDITOR

NOVINKA

New York

Library of Congress Cataloging-in-Publication Data

ISBN: 978-1-62417-907-5

Published by Nova Science Publishers, Inc. † New York

CONTENTS

PREFACE

This book examines the U.S.-foreign partnership efforts to prevent nuclear smuggling and terrorist travel. The National Nuclear Security Administration (NNSA) established the Megaports Initiative in 2003 to deter, detect, and interdict nuclear or other radiological materials smuggled through foreign seaports. The Initiative funds the installation of radiation detection equipment at select seaports overseas and trains foreign personnel to use this equipment to scan shipping containers entering and leaving these seaports, regardless of destination. Additionally, the U.S. government has identified four key gaps in foreign countries' capacity to prevent terrorist travel overseas. This book evaluates how U.S. capacity-building efforts address those gaps and assesses the extent to which the U.S. government is measuring progress in its efforts to close those gaps.

Chapter 1 – NNSA established the Megaports Initiative in 2003 to deter, detect, and interdict nuclear or other radiological materials smuggled through foreign seaports. The Initiative funds the installation of radiation detection equipment at select seaports overseas and trains foreign personnel to use this equipment to scan shipping containers entering and leaving these seaports—regardless of destination. NNSA provides partner countries with maintenance and technical support for about 3 years, after which it transfers the equipment and all related responsibilities to partner countries. GAO was asked to examine (1) the status of the Megaports Initiative and NNSA's plans for completing and sustaining it and (2) the benefits of the Initiative and factors that reduce its effectiveness. GAO analyzed key documents; interviewed agency officials; and visited eight Megaports in five countries, selected on the basis of port size and unique characteristics, among other things.

Chapter 2 – Eliminating the threat of terrorist attacks continues to be a primary U.S. national security focus. According to the 9/11 Commission, constraining the mobility of terrorists is one of the most effective weapons in fighting terrorism. This report (1) describes key gaps the U.S. government has identified in foreign countries' capacity to prevent terrorist travel overseas, (2) evaluates how U.S. capacity-building efforts address those gaps, and (3) assesses the extent to which the U.S. government is measuring progress in its efforts to close those gaps. To identify the key gaps, GAO reviewed governmentwide assessments of vulnerabilities in the international travel system. GAO reviewed the strategies and documentation of U.S. agencies funding and/or implementing foreign capacity-building efforts to prevent terrorist travel overseas, including those of the Departments of State (State)—which coordinates U.S. efforts overseas— Defense (DOD), Homeland Security (DHS), Justice (DOJ), and the U.S. Agency for International Development (USAID). GAO also interviewed officials from the National Security Staff, of the National Security Council (NSC), which oversees counterterrorism policy. GAO met with these agencies and conducted field work in Kenya, Pakistan, the Philippines, and Thailand.

In: Prevention of Nuclear Smuggling … ISBN: 978-1-62417-907-5
Editor: Valere Pruitt

Chapter 1

COMBATING NUCLEAR SMUGGLING: MEGAPORTS INITIATIVE FACES FUNDING AND SUSTAINABILITY CHALLENGES*

United States Government Accountability Office

WHY GAO DID THIS STUDY

NNSA established the Megaports Initiative in 2003 to deter, detect, and interdict nuclear or other radiological materials smuggled through foreign seaports. The Initiative funds the installation of radiation detection equipment at select seaports overseas and trains foreign personnel to use this equipment to scan shipping containers entering and leaving these seaports— regardless of destination. NNSA provides partner countries with maintenance and technical support for about 3 years, after which it transfers the equipment and all related responsibilities to partner countries. GAO was asked to examine (1) the status of the Megaports Initiative and NNSA's plans for completing and sustaining it and (2) the benefits of the Initiative and factors that reduce its effectiveness. GAO analyzed key documents; interviewed agency officials; and visited eight Megaports in five countries, selected on the basis of port size and unique characteristics, among other things.

---

* This is an edited, reformatted and augmented version of United States Government Accountability Office, Publication No. GAO-13-37, dated October 2012.

What GAO Recommends

GAO recommends that NNSA take actions, including (1) finalizing its longterm plan for ensuring the sustainability of Megaports operations after NNSA's final transfer of equipment to partner countries and (2) developing and maintaining useful and reliable measures to assess the performance of the Initiative. GAO also recommends that NNSA and DHS jointly assess the extent to which the two Initiatives are effectively coordinating. NNSA and DHS agreed with GAO's recommendations.

What GAO Found

As of August 2012, the National Nuclear Security Administration (NNSA) had completed 42 of 100 planned Megaports projects in 31 countries and, as of December 2011, NNSA had spent about $850 million on the Megaports Initiative (Initiative). NNSA's Initiative has equipped these seaports with radiation detection equipment, established training programs for foreign personnel, and created a sustainability program to help countries operate and maintain the equipment. However, the administration's fiscal year 2013 budget proposal would reduce the Initiative's budget by about 85 percent, and NNSA plans to shift the Initiative's focus from establishing new Megaports to sustaining existing ones. As a result, NNSA has suspended ongoing negotiations and cancelled planned deployments of equipment in five countries.

Officials from the five countries GAO visited reported benefits of the Megaports Initiative, including increased capacity to interdict nuclear and radiological materials. However, GAO identified several factors that reduce the Initiative's effectiveness. For example, NNSA has not finalized a long-term plan for ensuring the sustainability of Megaports operations after NNSA transfers radiation detection equipment to partner countries. Without a long-term plan for ensuring countries' ability to continue Megaports operations, NNSA cannot be assured that its $850 million investment will be sustained. Moreover, the Initiative's performance measures do not provide sufficient information for decision making because they do not evaluate the impact and effectiveness of the Initiative. GAO has previously reported that agencies successfully assess performance when they use measures that demonstrate results, cover multiple program priorities, and provide useful information for decision making. GAO also found that the Megaports Initiative and the Department of Homeland Security's (DHS) Container Security Initiative (CSI)—a related program that examines high-risk shipping containers for weapons of mass destruction before they are shipped to the United States—are not sufficiently coordinating. The two Initiatives are co-located at 29 foreign seaports. In two countries, DHS officials told GAO that they were using personal radiation detectors—a type of equipment intended for personal safety but not appropriate for scanning containers—to inspect containers if their radiation detection equipment is broken. However, in both countries, the Megaports Initiative had more suitable equipment that DHS officials could have used.

ABBREVIATIONS

ASP	Advanced Spectroscopic Portal
CSI	Container Security Initiative
DHS	Department of Homeland Security
DOE	Department of Energy
MRDIS	Mobile Radiation Detection and Identification System
NNSA	National Nuclear Security Administration
OMB	Office of Management and Budget
PNNL	Pacific Northwest National Laboratory
PRD	Personal Radiation Detector
RDSC	Radiation Detection Straddle Carrier
RIID	Radioisotope Identification Device
RPM	Radiation Portal Monitor
SLD	Second Line of Defense
SPM	Spectroscopic Portal Monitor

October 31, 2012

The Honorable Elijah Cummings
Ranking Member
Committee on Oversight and Government Reform
House of Representatives

The Honorable Edolphus Towns
House of Representatives

Over the past decade, the United States has become increasingly concerned about the threat posed by unsecured weapon-usable nuclear material,[1] which could be stolen and fall into the hands of terrorists or countries seeking weapons of mass destruction. According to the International Atomic Energy Agency, from 1993 to 2011, there were 2,164 confirmed incidents of illicit trafficking in nuclear and radiological materials. Even small amounts of nuclear materials are of concern—as little as 25 kilograms of highly enriched uranium or 8 kilograms of plutonium could be placed inside containers aboard cargo ships and ultimately be used to build a nuclear weapon known as an improvised nuclear device. An improvised nuclear device could create an explosion producing extreme heat, powerful shock waves, and intense radiation that would be immediately lethal to individuals within miles of the explosion, as well as radioactive fallout over thousands of square miles. Radiological materials, such as cobalt-60, cesium-137, and strontium-90, also pose a significant security threat. These materials are used worldwide for many legitimate medical and industrial purposes. However, they could also be used by a terrorist to produce a simple and crude, but potentially dangerous weapon, known as a radiological dispersion device, or dirty bomb. Experts believe that a dirty bomb detonation could result in a limited number of deaths but that it could have severe economic, psychological, and social consequences.

Seaports are critical gateways for international commerce, and maritime shipping containers play a vital role in the movement of cargo between global trading partners. In 2009, for example, world maritime container traffic (loaded and empty) was estimated at over 432 million twenty-foot equivalent units.[2] The Department of Energy's (DOE) National Nuclear Security Administration (NNSA)[3] established the Megaports Initiative in 2003 so that foreign personnel at select seaports could use radiation detection equipment to scan shipping containers entering and leaving these seaports—regardless of the containers' destination—for nuclear or radiological material. The Megaports Initiative (Initiative) is implemented by NNSA's Office of the Second Line of Defense (SLD), which aims to strengthen foreign countries' capability to deter, detect, and interdict illicit smuggling of nuclear and radioactive materials across international borders. Through the Initiative, NNSA partners with foreign governments to fund the installation of radiation detection equipment at foreign seaports and train foreign personnel to operate this equipment. In some cases, partner countries also contribute funding to the design, construction, or installation of radiation detection equipment. The installed radiation detection equipment is then operated by foreign personnel.

In addition, the Initiative provides maintenance and technical support for the equipment for about 3 years, after which NNSA transfers all responsibilities—including operations and maintenance of the equipment and all related financial responsibilities—to partner countries. NNSA selects seaports for participation in the Initiative using a prioritization model that ranks seaports on the basis of scannable shipping volume and threat factors, such as the origin and destination of container movement through a port and proximity to special nuclear material.[4]

We reported in March 2005 that the Megaports Initiative had limited success in initiating work at seaports identified as high priority.[5] Among other things, we reported that it was difficult to gain the cooperation of foreign governments, in part because some countries were concerned that scanning large volumes of containers would create delays, thereby inhibiting the flow of commerce at their ports. Moreover, the Initiative did not have a comprehensive long-term plan to guide the Initiative's efforts and faced several operational and technical challenges in installing radiation detection equipment at foreign seaports. We recommended that DOE (1) develop a comprehensive long-term plan for the Initiative that identifies criteria for deciding how to strategically set priorities for establishing Megaports and (2) reevaluate cost estimates and adjust long-term projections as necessary. DOE has implemented both recommendations.

In this context, you asked us to update information on the Megaports Initiative since our 2005 report. Accordingly, we examined (1) the status of the Megaports Initiative and NNSA's plans for completing and sustaining it and (2) the benefits of the Initiative and factors, if any, that reduce its effectiveness.

For both objectives, we obtained and analyzed relevant NNSA documentation, including program plans and implementation strategies, memorandums of understanding, and cost-sharing arrangements between NNSA and partner countries, as well as sustainability plans. We interviewed NNSA officials about the status of the program, its future plans, and challenges associated with its implementation. Moreover, we interviewed contractors who help implement the Initiative at two DOE national laboratories—Pacific Northwest National Laboratory (PNNL) and Oak Ridge National Laboratory. In addition, we visited a nonprobability sample of eight seaports in five countries[6]—Belgium, the Dominican Republic, Jamaica, Panama, and Spain—to observe scanning operations and interview foreign government and port officials.[7] We selected these seaports on the basis of port size, NNSA's priority ranking, NNSA's expenditures at the port, history of cost-sharing with NNSA, length of time in the Megaports Initiative, joint presence with the Department of Homeland Security's (DHS) Container Security Initiative (CSI),[8] and unique characteristics of specific locations. Because we used a nonprobability sample to select the eight Megaports we visited, the observations we made are not generalizeable to other Megaports but serve as examples of challenges and concerns that may affect Megaports operations and sustainability. At the seaports visited, we interviewed foreign government officials and terminal operators using a standard set of questions about their participation in and perspective on the Megaports Initiative. We also interviewed DHS officials co-located at these seaports and U.S. embassy officials in these countries.

To assess the status of the Megaports Initiative and NNSA's plans for completing and sustaining it, we analyzed the Megaports Initiative's budget information and expenditures for fiscal years 2003 through 2012, as well as projected costs through fiscal year 2018. We interviewed knowledgeable NNSA officials to assess the reliability of the data, including the issues of data entry, access, quality control procedures, and the accuracy and completeness of the data. We determined that the data were sufficiently reliable for our purposes. We also requested and analyzed data on Megaports partner countries' cost-sharing contributions. These cost-sharing data are of undetermined reliability because we did not have a way to verify NNSA's and partner countries' estimates. We also reviewed NNSA's budget request and related budget documents for fiscal year 2013 and spoke with Office of Management and Budget (OMB) staff about proposed changes to the Megaports Initiative's fiscal year 2013 budget. To assess the benefits of the Initiative and any factors that may reduce its effectiveness, we also interviewed NNSA headquarters, PNNL, and foreign officials about the benefits and challenges they have encountered in implementing the Initiative. Additional details on our objective, scope, and methodology can be found in appendix I.

We conducted this performance audit from August 2011 to October 2012 in accordance with generally accepted government auditing standards. Those standards require that we plan and perform the audit to obtain sufficient, appropriate evidence to provide a reasonable basis for our findings and conclusions based on our audit objectives. We believe that the evidence obtained provides a reasonable basis for our findings and conclusions based on our audit objectives.

BACKGROUND

SLD manages three programs: the Megaports Initiative, the Core Program, and the Sustainability Program. The Core Program funds the installation of radiation detection equipment at land borders, airports, and strategic seaports—primarily in Russia, other former Soviet Union countries, and Eastern Europe. The Sustainability Program supports both the Megaports Initiative and the Core Program by overseeing the transition of radiation detection equipment from NNSA to partner countries. Figure 1 shows the organization of SLD.

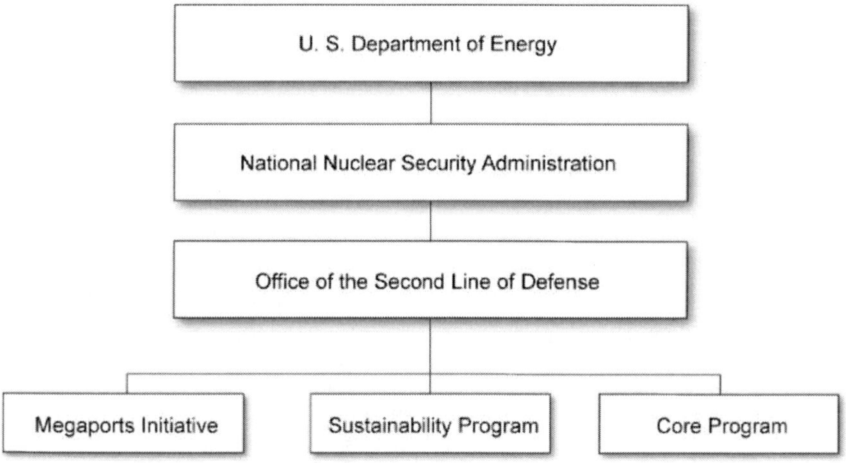

Source: GAO analysis of NNSA documents.

Figure 1. Organizational Structure of SLD.

The Megaports Initiative is part of a broader U.S. government effort to prevent terrorists from acquiring, smuggling, and using special nuclear and other radioactive materials to develop a weapon of mass destruction or a dirty bomb. The Megaports Initiative coordinates with and complements DHS's Container Security Initiative (CSI), a related program that targets and examines high-risk containers for weapons of mass destruction before they are shipped to the United States. CSI began operating in January 2002 and is active at 58 seaports around the world, including 29 seaports where the Megaports Initiative had completed Megaports as of August 2012.

Table 1. Key Attributes of the Megaports and Container Security Initiatives

Program	Lead U.S. agency	Mission	Personnel involved	Containers scanned	Equipment used	Goal
Megaports Initiative	NNSA	NNSA funds or cost-shares the installation of radiation detection systems at select foreign seaports and trains foreign personnel to scan shipping containers for nuclear and other radioactive materials.	Foreign personnel trained by the Megaports Initiative	As many containers as possible, regardless of destination	Radiation detection equipment: radiation portal monitors and handheld detection equipment	Detect nuclear or other radioactive material
Container Security Initiative	DHS	U.S. Customs and Border Protection officials posted at foreign seaports identify high-risk shipping containers and request that foreign authorities examine the containers with nonintrusive imaging systems or physical inspections before they are loaded on vessels destined for the United States.	U.S. Customs and Border Protection officials and foreign personnel	Selected U.S.-bound containers only	Nonintrusive imaging systems and handheld radiation detection equipment	Detect potentially dangerous cargo,such as weaponsof mass destruction and illicit drugs

Sources: GAO analysis of NNSA and DHS documents.

Under CSI, DHS places U.S. Customs and Border Protection officials at foreign seaports to review the cargo manifests of U.S.-bound shipping containers and identify high-risk containers for scanning with nonintrusive imaging systems—which use X-rays or gamma rays to scan a container and create images of the container's contents without opening it. In some cases, CSI officers use handheld radiation detection equipment to collect additional information about the selected shipping containers. The Megaports Initiative and the CSI programs have some similarities, but they also have key differences, which are outlined in table 1.

Table 2. Roles of SLD, DOE National Laboratories, and Private Contractors Involved in the Megaports Initiative

Responsible Megaports entity	Role
NNSA's SLD	Develops the overall strategic direction and identifies the priority goals and objectives for the Initiative. Oversees and manages all aspects of Megaports implementation through the entire project life cycle.
Pacific Northwest National Laboratory	Provides project management, training, and sustainability support, and conducts some equipment procurements. Project managers oversee implementation and sustainability activities, including the scope, schedule, and budget of Megaports.
Sandia National Laboratories	Conduct site and communications surveys for new Megaports. Prepare design and communications requirements for new Megaports. Conducts testing on the communications systems that will be deployed to Megaports and leads in-country testing of entire radiation detection systems.
Los Alamos National Laboratory	Conducts background radiation survey at new Megaports. Makes technical recommendations on settings for radiation detection equipment. Conducts equipment testing at completed Megaports. Conducts laboratory-based testing on equipment and new and emerging technologies to determine detection capabilities and performance.
Oak Ridge National Laboratory	Conducts a monthly analysis of data from radiation detection equipment from operational Megaports to ensure that the equipmentis functioning properly.
Small-Business U.S. Contractors[a]	Provide design, integration, construction, communication, and engineering expertise to support deployment of new Megaports.

Source: GAO analysis of NNSA documents.

[a]For each new Megaport, NNSA selects one of three preapproved contractors using a competitive bid process.

The *Megaports Initiative 2011 Program Plan* states that the Initiative has committed to providing radiation detection capability at all CSI ports, which will give CSI officials additional information in assessing high-risk containers. SLD, numerous DOE national laboratories,[9] and private contractors each play a role in implementing the Megaports Initiative, as described in table 2.

The Megaports team generally implements the Initiative at foreign seaports in three phases: (1) engagement, (2) implementation, and (3) sustainability.

- *Engagement*: Before installing equipment at a seaport, NNSA and the appropriate agency in the partner country enter into an arrangement that outlines the expectations, roles, and responsibilities of all stakeholders involved. When possible, NNSA attempts to establish a cost-sharing arrangement with partner countries.
- *Implementation*: This phase involves several different activities. The Megaports team surveys the port to help plan for the installation of radiation detection equipment. Moreover, the Megaports team begins, among other things, training foreign officials on operating radiation detection equipment, resolving alarms, and maintaining this equipment. During this phase, the Megaports team also (1) hires one of three approved private contractors that will lead the construction and communications services at the new Megaport, (2) drafts a sustainability and a joint transition plan, and (3) installs and tests the radiation detection equipment and prepares the equipment for operations.
- *Sustainability*: The Megaports team works with the partner country to develop the capabilities needed to sustain equipment operations and maintenance over the long-term. For example, the Megaports Initiative funds the maintenance of equipment during this phase and conducts additional training of foreign officials. NNSA officials also regularly visit the port to ensure that the equipment is properly calibrated and being used as intended. NNSA typically transfers ownership of and responsibility for radiation detection equipment operations and maintenance to partner countries after about 3 years.

The Megaports Initiative provides partner countries with several types of radiation detection equipment that is used to scan shipping containers in primary or secondary scanning inspections. NNSA's goal is to scan as much container traffic as possible with primary radiation detection equipment—

including Radiation Portal Monitors (RPM), which are large stationary systems through which cargo containers and trucks pass as they are entering or exiting the port; Mobile Radiation Detection and Identification Systems (MRDIS), which are mobile radiation detection systems that can be driven to different locations at a seaport; and Radiation Detection Straddle Carriers (RDSC), which are mobile radiation detection systems that can scan rows of containers stacked up to three high. These types of equipment detect both gamma and neutron radiation, which is important for detecting the presence of highly enriched uranium and plutonium. When a container is scanned with primary equipment, an alarm will sound if the equipment detects the presence of radiation. However, during primary scanning, the equipment cannot identify what type of material triggered the alarm. If they determine it is necessary, foreign customs officials will then conduct a secondary inspection with a handheld Radioisotope Identification Device (RIID) or a Spectroscopic Portal Monitor (SPM) to specifically identify the source and location of the radiation. If the foreign customs officials cannot determine the content of the container after the secondary inspection, they may manually inspect the container or request assistance from other agencies within their government. NNSA also provides partner countries with Personal Radiation Detectors (PRD), or pagers, which are small detectors that can be worn on an individual's belt to continuously monitor radiation levels in the immediate area. PRDs help to ensure personal safety. For more detailed information about and photographs of the equipment NNSA provides partner countries, see appendix II.

NNSA Plans to Complete 45 Megaports, but Future Implementation Is Uncertain Because of Administration-Proposed Budget Cuts

NNSA completed 42 Megaports as of August 2012 and plans to complete work at 3 additional ports by the end of fiscal year 2012, for a total of 45 Megaports at a cost of about $850 million.[10] However, if proposed budget cuts are enacted, the Initiative's plans to establish new Megaports would cease. As a result, NNSA has halted the implementation of new Megaports and is working with an interagency work group to determine an appropriate future scope for the program.

NNSA Plans to Complete 45 Megaports and Has Spent about $850 Million

NNSA had completed 42 Megaports as of August 2012 and plans to complete 3 additional ports by September 2012, according to NNSA officials, bringing the total to 45 operational Megaports by the end of fiscal year 2012. To ensure that countries are able to independently operate and maintain the radiation detection equipment after NNSA transfers the equipment, NNSA has (1) implemented a training program for foreign personnel operating the radiation detection equipment, (2) developed a sustainability program, and (3) developed new technology to address challenges faced at some ports. As of December 2011, NNSA had spent about $850 million on Megaports Initiative activities.

Status of the Megaports Initiative

The 42 completed Megaports and the 3 additional Megaports that are planned for completion by September 2012 are located in 33 countries throughout Europe, Asia, Latin America, Africa, and the Middle East. Figure 2 shows the countries where NNSA has either established a Megaport or plans to have completed work for one by the end of fiscal year 2012. (See app. III for a complete list of the Megaports and information about each port.)

Instructions: Online, hover over the country names in the graphic for more information. For print version, see appendix IV.

To help decision makers identify and prioritize foreign seaports for participation in the Megaports Initiative, NNSA uses a model that ranks foreign ports according to their relative attractiveness to potential nuclear smugglers. Currently, the model scores 1,100 foreign seaports on the basis of two categories: (1) scannable shipping volume,[11] which accounts for 75 percent of the score, and (2) potential threat, which accounts for 25 percent of the score. Potential threat is determined on the basis of several factors, including the capabilities of terrorist groups within a country, amount of special nuclear material within a country, and the freedom of criminal groups to operate within a country. This information is then combined to provide each port with an overall score. Ports receiving higher scores are considered more attractive to a nuclear material smuggler and, therefore, of potentially higher interest for inclusion in the Initiative.

The model is also updated regularly to incorporate new information. After the model produces a ranked list of ports that takes into consideration these factors and assumptions, NNSA officials work with the Department of State to ensure that U.S. diplomatic concerns are considered. For example, NNSA may not pursue ports in countries that do not have diplomatic relations with the United States.

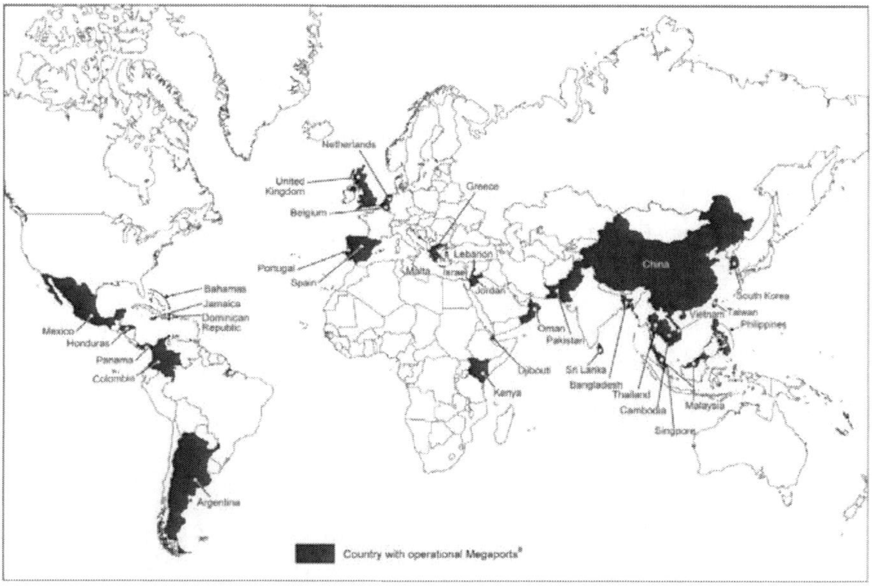

Source: DOE; Map Resources (map).

[a] DOE expects the following ports to be operational as of September 2012: (1) Buenos Aires, Argentina, (2) Port of Cebu, Philippines, (3) Cai Map, Vietnam.

Figure 2. Countries with Operational Megaports.

According to our analysis of NNSA's priority list of ports, 18 of the 45 Megaports that are completed or under way (or 40 percent) are ranked lower in priority than the top 100 highest priority seaports NNSA identified. NNSA officials told us that the Initiative uses the priority list generated by the model as a targeting guide but noted that various factors affect which ports they can equip with radiation detection equipment. For example, some foreign governments with seaports identified as being of high-priority have been unwilling to participate in the Initiative because of internal political sensitivities. NNSA officials also told us that the DHS CSI ports are always added to the list regardless of ranking because of an arrangement between NNSA and DHS for the Megaports Initiative to provide radiation detection equipment at these ports. In addition, NNSA officials said that they choose some lower ranked ports if there are cost-sharing opportunities, if partner countries specifically request partnerships, or if NNSA is given an opportunity to demonstrate a new scanning approach—such as using the MRDIS to scan transshipped containers in Salalah, Oman.

We last reported on the Megaports Initiative in March 2005. At that time, NNSA had completed work on two Megaports and faced diplomatic and other impediments that limited its ability to complete more Megaports in its first 2 years.[12] In addition, we reported that NNSA officials faced difficulties reaching agreement with foreign countries due to political considerations outside of NNSA's control. In October 2011, NNSA officials told us they continue to face these same limitations. For example, NNSA officials said that they had been negotiating with one country's government on installing a Megaport when one of that country's citizens was killed by U.S. armed forces in Iraq. According to NNSA officials, this incident prompted that country's government to cease negotiations. After 9 months, negotiations resumed, but the progress previously made was lost, and NNSA and the officials in that country had to start negotiations over.

The Megaports Initiative Has Implemented a Training and Sustainability Program and Developed New Technology

To ensure that countries are able to independently operate and maintain the radiation detection equipment after NNSA transfers it, PNNL implements a training program for foreign personnel operating the radiation detection equipment. PNNL officials told us that, as of December 2011, they had trained 1,038 foreign officials at DOE's Hazardous Materials Management and Emergency Response training facility in Washington State, as well as 1,521 foreign officials in their respective countries. These training courses include instruction on, among other things, operating radiation detection equipment at ports, responding to alarms generated by radiation detection equipment, using handheld instruments for secondary inspections, and maintaining the radiation detection equipment.

Through the SLD Sustainability Program, NNSA works with foreign partners to develop a joint transition and a sustainability plan that lays out actions needed at the port before NNSA transfers the equipment. These plans may include, among other things, steps for scheduled maintenance and refresher training. NNSA also provides ongoing support for the Megaports equipment through the SLD help desk managed by PNNL. The help desk provides partner countries with access to technical expertise—even after the equipment is transferred—so that they can resolve problems they may experience with the radiation detection equipment.

The Megaports Initiative has developed new technologies to address transshipment challenges, as well as to improve overall scanning effectiveness and minimize the impact on seaport operations. According to the Initiative's program plan, transshipped cargo presents a significant challenge to a port's ability to scan containers because of the (1) shorter transfer time for those containers to be at the seaport before they are moved to another location or port, (2) space constraints, (3) lack of shipping data, and (4) difficulty of identifying fixed locations to place RPMs within terminals that do not impede seaport operations. To address these challenges at seaports with high transshipment volumes, NNSA developed the MRDIS and the RDSC, which are two types of mobile radiation detection equipment that can be driven to different locations within a seaport (see figs. 3 and 4). Appendix II provides additional details on these and other types of equipment used by the Megaports Initiative.

Source: GAO.

Figure 3. Truck Driving through a Mobile Radiation Detection and Identification System.

Source: GAO.

Figure 4. Radiation Detection Straddle Carrier Scanning Shipping Containers.

Megaports Initiative Expenditures

As of December 2011, NNSA had spent about $850 million on Megaports Initiative activities, according to our analysis of NNSA expenditures. Of this total, NNSA spent 64 percent, or about $547 million, in countries where it has installed radiation detection equipment and 36 percent, or about $303 million, on expenditures related to program integration activities, secondary equipment, and outreach to countries. (See app. V for information on the Initiative's total budget and expenditures for fiscal years 2003 through 2012.) Program integration activities include expenditures for project management, technical support, equipment testing, and equipment storage. NNSA officials told us the costs of installing equipment at individual seaports vary and are influenced by factors such as a seaport's size, its physical layout, existing infrastructure, and the type of radiation detection equipment deployed. Table 3 shows the Initiative's expenditures by country from fiscal year 2003 through December 2011. We found that the amount of money NNSA spent per country ranged from about $1 million to $59 million, with an average of $16.6 million.

Table 3. Megaports Initiative's Expenditures by Country, Fiscal Year 2003 through December 2011

Constant dollars in thousands		
Country	**Number of operational Megaports**	**NNSA expenditures**
Belgium	2	$59,307
Taiwan	1	51,973
Malaysia	2	41,511
Spain	3	34,238
Philippines[a]	2	29,436
Sri Lanka	1	28,188
Bahamas	1	25,500
Mexico	4	25,365
Thailand	1	19,830
Oman	1	19,154
China	1	19,078
Jordan[b]	1	16,879
Panama	4	16,508
Kenya	1	14,476
Bangladesh	1	13,010
United Kingdom	1	12,811
Israel	2	12,298
South Korea	1	11,571
Portugal	1	10,018
Jamaica	1	10,000
Djibouti	1	9,727
Lebanon	1	9,549
Singapore	1	9,513
Argentin[aa]	1	8,801
Dominican Republic	1	8,290
Honduras	1	6,082
Colombia	1	5,287
Cambodia	1	5,121
Netherlands	1	4,252
Pakistan	1	3,025
Greece	1	2,988
Vietnam[a]	1	2,202
Malta	1	1,248
Additional costs[c]		302,550
Total	**45**	**$849,784**[d]

Source: GAO analysis of NNSA documents.

[a]DOE expects a new Megaport in each of these three countries to be operational as of September 2012.

[b]The Megaports Initiative installed equipment at four different locations in Jordan, including at one port in Aqaba and at three land border sites. NNSA considers all work in Jordan as one Megaport, and the costs listed for Jordan include the costs for all four sites.

[c]Additional costs include costs related to program integration activities ($267.7 million), secondary equipment (about $19.7 million), and outreach to countries (about $15.2 million). Program integration activities include expenditures for project management, training, and equipment testing. [d]Amounts may not total due to rounding.

Partner Countries' Cost-Sharing Contributions

According to NNSA data, eight foreign countries have contributed an estimated $44 million to support Megaports at 15 foreign ports (see table 4 for a list of these countries and their estimated contributions). According to NNSA, partner country cost-sharing estimates are derived from educated estimates and partner country-provided data.[13] Cost-sharing arrangements typically include the partner country funding the design, construction, and installation of the radiation detection equipment and NNSA providing equipment, training, and technical and maintenance support. According to the *Megaports Initiative 2011 Program Plan*, NNSA employs a formal cost-sharing approach with a partner country whenever possible. Moreover, according to the Initiative's program plan, when a partner country contributes a percentage of the cost and work to install the equipment, the country develops a greater sense of ownership for the program and contributes to the Initiative's success.

NNSA's cost-sharing estimates do not, however, include informal contributions by countries or other foreign partners, such as seaport terminal operators. Both NNSA officials and foreign partners stated that foreign partners have made contributions that were not specified in a cost-sharing arrangement. In reviewing NNSA documents, we identified at least nine instances where there were no formal cost-sharing arrangements in place, but NNSA had acknowledged informal contributions from foreign partners that could be identified as monetary savings for the Initiative. For example, the Dominican Republic does not have a formal cost-sharing arrangement in place, but the terminal operator contributed fiber optic cables for the Megaports Initiative at the port of Caucedo, which, according to NNSA officials, contributed to monetary savings. In another case, a terminal operator in the Philippines provided power and the infrastructure for fiber optic cables to support the Initiative, which ultimately saved the Initiative both financial resources and construction time.

In December 2011, we reported that NNSA does not systematically track and verify foreign cost sharing for the Megaports Initiative and that a NNSA official told us that the Megaports Initiative cost-sharing values are estimates and are not validated or audited.[14] In that report, we recommended that NNSA develop ways to better track and maintain information on foreign cost sharing for nuclear nonproliferation projects overseas. NNSA neither agreed nor disagreed with this recommendation.

Table 4. Estimated Partner Country Cost-sharing Contributions

Dollars in thousands			
Country	Megaport	NNSA expenditures[a]	Estimated total partner country contributions[b]
Belgium	Antwerp	$51,448	$10,363
Belgium	Zeebrugge	7,859	7,000
China	Shanghai	19,078	256
Colombia	Cartagena	5,287	325
Israel	Ashdod	5,847	400
Israel	Haifa	6,450	1,240
Mexico	Altamira	3,909	2,400
Mexico	Lazaro Cardenas	5,268	7,100
Mexico	Manzanillo	8,713	9,560
Mexico	Veracruz	7,475	375
Panama	Manzanillo International Terminal	3,223	73
Singapore	Singapore	9,513	160
Spain	Algeciras	19,882	1,500
Spain	Barcelona	6,808	1,150
Spain	Valencia	7,548	2,170
Total		$168,308	$44,072

Source: GAO analysis of NNSA documents.

[a] Expenditures are from fiscal year 2003 through December 2011. Expenditures include training, project management, and sustainability. NNSA expenditures are in constant dollars.

[b] Cost-sharing data are from fiscal year 2003 through February 2012. Contributions listed are for those countries with formal cost-sharing arrangements with NNSA. Country contributions are not adjusted for inflation because the years of expenditures were not known.

As the table shows, Mexico, Belgium, and Spain shared the most costs with the Megaports Initiative, contributing about $19 million, $17 million, and $5 million, respectively.

Conversely, Panama and China contributed the least amount of funding: $72,600 and $256,000, respectively. In January 2007, we reported that nuclear nonproliferation is a shared responsibility and that some countries should be willing to contribute more resources to secure nuclear and radiological materials.[15]

We recommended that DOE develop strategies to encourage cost sharing with recipient countries. DOE stated that the recommendation was very helpful and would further strengthen its program.

Due to Proposed Budget Cuts, NNSA Plans to Shift the Initiative's Focus to Sustainment Rather than Expansion

NNSA had planned to establish a total of 100 Megaports by 2018, which NNSA officials told us would cost an additional $1.26 billion.[16] These plans have been halted, however, because the administration's fiscal year 2013 budget proposal would reduce the SLD program budget from $262.1 million to $92.6 million, of which $19.6 million would be allocated to the Megaports Initiative.[17] This is an 85 percent reduction for the Megaports Initiative—which had a $132.7 million budget in fiscal year 2012.[18] As a result, DOE's fiscal year 2013 Congressional Budget Request states that DOE plans to shift the program's focus from establishing new Megaports to sustaining existing ones. See table 5 for a summary of SLD's budget for fiscal year 2012 and budget request for fiscal year 2013.

Table 5. SLD Budget for Fiscal Year 2012 and Budget Request for Fiscal Year 2013

Current dollars in thousands			
	Fiscal year 2012 budget enacted	Fiscal year 2013 budget request	Percentage decrease
Megaports	$132,670	$19,574	85.2
Core Program	$129,402	$73,000	43.6
Total (SLD Program)	$262,072	$92,574	64.7

Source: GAO analysis of NNSA documents.

NNSA has taken a number of actions as a result of proposed budget cuts, including suspending ongoing negotiations for installing Megaports in 17 countries and cancelling planned deployments of new equipment in 5 other countries. NNSA officials told us that arrangements with 2 of these countries were imminent, and the Initiative's former program manager stated that cancelling program activities would affect the agency's credibility. According to NNSA officials, it would also significantly diminish countries' trust in working with NNSA in the future. Moreover, NNSA typically funds maintenance of Megaports in transition on a yearly basis, but because of the expected budget cuts, NNSA has decided to fund maintenance contracts up to 3 years in advance of transferring the radiation detection equipment for some countries in order to honor its commitments.

The Office of Management and Budget (OMB) staff,[19] who examined the SLD budget as part of their role in preparing the administration's budget, cited several reasons for reducing the Megaports budget, including the following:

- *Diminishing impact*: The Megaports Initiative generally follows a prioritized scope of work for installing radiation detection equipment and has already equipped many of the world's most important seaports. As the Megaports Initiative completes the installation of new Megaports, the benefit of any significant expansion, except at certain priority sites, diminishes.
- *Questionable program effectiveness*: The technologies used in radiation detection equipment may have some limitations because they may not be able to detect nuclear or other radiological material that has been shielded or masked, and terrorists could also bypass the Megaports Initiative's equipment.[20]
- *Lack of performance measures*: It is difficult to measure the Megaports Initiative's progress in achieving its mission using its existing performance measures. Specifically, the program has limited performance measures—the total number of foreign seaports where NNSA has installed radiation detection equipment and the cumulative number of Megaports with partner country cost-sharing. Moreover, NNSA does not report on what has been interdicted as a result of the Megaports Initiative, although it does internally collect this information from willing partner countries. OMB staff said that reporting the number of interdictions and what has been interdicted would be useful to help assess the Initiative's effectiveness in the future.
- *Limited monitoring of equipment*: It is difficult for NNSA to monitor how partner countries, as sovereign nations, are using the radiation detection equipment once it transfers the equipment to a partner country.
- *Potential duplication*: The federal government has 21 nonproliferation programs, including the Megaports Initiative. Because of the current fiscally constrained environment, OMB staff examined national security programs and identified priorities for funding according to which programs are most effective in combating nuclear smuggling. According to OMB staff, within the fiscal constraints of the *Budget Control Act of 2011*, this strategic prioritization enabled the administration to protect its highest priority nonproliferation

activities, most importantly those that directly secure or remove material.

Notwithstanding the reasons they identified for reducing the Initiative's budget, OMB staff told us that there are benefits to the Megaports Initiative, such as providing a public health benefit. For example, they said that using the Megaports radiation detection equipment, countries have interdicted scrap metal contaminated with radiological material, thereby preventing it from circulating within the country or beyond its borders. Additionally, OMB staff told us that the Initiative helps deter terrorists from smuggling nuclear materials through the global shipping network. Furthermore, the Initiative has helped to build relationships with partner countries and has demonstrated the U.S. commitment to combating nuclear terrorism. OMB staff said that they weighed the program's benefits and potential repercussions from foreign partners resulting from NNSA's cancellation of arrangements. However, they determined that the Initiative could not demonstrate that further expansion would have a significant impact on preventing nuclear or radiological materials from entering the United States.

Interagency Group Has Been Established to Review Megaports Initiative's Future Role

The administration's fiscal year 2013 budget proposes that NNSA, in coordination with members of an interagency work group, conduct a strategic review of SLD—including the Megaports Initiative—to determine an appropriate future scope for the Initiative and how it may affect other national security programs. NNSA officials told us that the interagency work group is led by the White House National Security Staff and includes officials from DHS's Domestic Nuclear Detection Office, the Department of State, the Department of Defense, and the National Counterterrorism Center. According to NNSA officials and OMB staff, NNSA is continuing its strategic review of the Megaports Initiative to ensure that the Initiative, under a constrained budget environment, is most effectively and efficiently detecting, deterring, and interdicting nuclear and radiological material. Through its strategic review, NNSA is assessing how to best leverage and sustain existing capabilities, and whether installing additional radiation detection systems at strategically selected ports is warranted to further strengthen detection and deterrence at key hubs and spokes in the global maritime shipping network.

The strategic review is assessing the threat of terrorists shipping nuclear materials through the maritime shipping network. NNSA's strategic review also includes an assessment of the Megaports Initiative's port prioritization methodology, in which NNSA updated its model with 2011 data. To generate a list of ports that would be realistic for consideration if expansion is warranted, NNSA is also considering additional factors, including a country's likelihood of entering into an arrangement, the type of equipment appropriate for each port, cost-sharing arrangements, and whether a country participates in other nuclear nonproliferation programs. According to NNSA's strategic review, through technical exchanges, the Megaports Initiative could provide assistance with technical aspects of the project involving the transfer of knowledge only. The results of the strategic review will be reflected in the fiscal year 2014 budget.

COUNTRIES CITE BENEFITS OF THE MEGAPORTS INITIATIVE, BUT SEVERAL FACTORS LIMIT ITS EFFECTIVENESS

Foreign officials from countries that we visited reported benefits of the Megaports Initiative, such as gaining a greater capacity to interdict radiological materials, but we also identified several factors that reduce the Initiative's effectiveness. These factors include partner countries that are not fully using radiation detection equipment or fully participating in the Initiative. The program also faces several management weaknesses, including having insufficient performance measures and limited testing of spectroscopic portal monitors and deployment of radiation detection equipment.

Benefits of the Megaports Initiative Cited by Partner Countries

Foreign officials from the five countries we visited highlighted several benefits of the Megaports Initiative. For example, foreign government officials from four countries—Belgium, the Dominican Republic, Jamaica, and Spain—told us that the Initiative has helped their respective countries interdict radiological materials found in containers using the Megaports radiation detection equipment. For example, Spanish customs officials told us that they interdicted medical isotopes that had not been properly disposed of, and

Jamaican officials told us that they interdicted contaminated scrap metal.[21] Moreover, officials from these four countries also told us that the radiation scanning equipment was used to ensure that goods imported from Japan were not contaminated in the wake of the 2011 Fukushima Daiichi nuclear incident. For example, Belgian customs officials told us that they interdicted 15 containers contaminated with radioactive material originating from Japan. In addition, according to foreign government officials, as part of the Initiative, NNSA officials helped both Jamaica and Spain develop a response plan in the event that their officials interdict nuclear or radiological material. Jamaican officials told us that seven different government agencies play a role in responding to a radioactive emergency and that the centralized response plan will help them better coordinate a response if they interdict radiological materials. NNSA officials also identified other program benefits. For example, according to NNSA officials, 13 countries with Megaports have reported about 155 interdictions of radiological materials. In addition, 19 countries have taken steps to supplement, or have expressed interest in supplementing, the Megaports Initiative by acquiring their own radiation detection equipment. According to the Initiative's program plan, based on the success of Megaports Initiative in the Netherlands, the Dutch government replaced four U.S.-provided RPM units with more than 40 RPMs at that country's expense to be able to scan all imports and exports at the Port of Rotterdam—which is the largest seaport in Europe. Spanish customs officials also told us that the Initiative provided a foundation for beginning radiation detection scanning in Spain. Spanish customs purchased handheld radiation detection equipment to use in airports and scanning equipment for six other seaports after the Fukushima accident.

Several Factors Have Impeded the Initiative's Effectiveness

Notwithstanding these benefits, we identified several factors that reduce the Initiative's effectiveness. Specifically, some partner countries are not fully using radiation detection equipment, and some foreign partners are not fully participating in the Initiative.

Some Partner Countries Not Fully Using Radiation Detection Equipment

In four of the five countries we visited—and seven of the eight seaports—countries were using the Megaports-funded secondary radiation detection equipment on a limited basis, and foreign officials told us that they are conducting secondary scanning for a small percentage of the containers that triggered an initial alarm. For example, customs officials from one country that we visited told us that they limit the containers that they refer for secondary scanning because they want to minimize interruptions to port operations. According to these officials, the day before we visited in January 2012, they referred one container for secondary scanning out of 17 containers that triggered alarms during primary scanning. Customs officials from this country select containers for secondary scanning by comparing the initial alarm information with a container's shipping information. For example, the shipping information should indicate what cargo the container holds, and on the basis of the information, officials assess whether that cargo naturally emits radiation.[22] If the shipping information matches the alarm information, they allow the container to proceed. Similarly, customs officials in another country we visited stated that if a container triggers a primary alarm, they compare the container's shipping information with the alarm information to determine whether the container should be scanned by secondary equipment. According to the *Megaports Initiative 2011 Program Plan*, one port in this country scans about 73 percent of containers with primary equipment; however, foreign officials from this port said that they conduct secondary scanning for only a small number of containers—about 2 to 3 containers per day out of about 200 primary alarms. (See fig. 5 for a photograph of a SPM radiation detection scanning unit, which is used for secondary scanning.)

Source: GAO.

Figure 5. A Spectroscopic Portal Monitor Used for Secondary Radiation Detection Scanning.

The Megaports Initiative encourages partner countries to conduct as many secondary inspections as possible, according to NNSA officials. However, NNSA officials also told us that they recognize that there are practical challenges to conducting secondary inspections on 100 percent of containers that trigger a primary alarm. For example, the flow of commerce may be affected when a truck is diverted from its normal route to a secondary inspection area or because of the time that it takes to conduct a secondary inspection. As a result, NNSA trains partner countries to set priorities for selecting containers for secondary inspection by assessing a container's shipping information, including, among other things, the commodity being shipped, the shipper, and the country of origin and destination.

In contrast, at U.S. ports of entry, 100 percent of containers that produce an initial alarm must be verified with more precise secondary scanning equipment, according to DHS guidance. DHS officials, who manage the domestic radiation detection program, told us that secondary scanning is important to reduce the risk that containers may be masking or shielding nuclear material. In addition, the *Megaports Initiative 2011 Program Plan* states that, although RPMs are the most sensitive radiation detection technology deployed by the Megaports Initiative, they do not have the ability to unambiguously characterize the location, distribution, and identity of a radiation source in a container that triggers an alarm. The program plan also states that this type of information is determined by a secondary inspection. Furthermore, a senior official for a global terminal operator told us that it is "very easy" for exporters to lie about the contents of a shipping container. According to this official, smugglers could also place materials into a container after it leaves an exporter's facilities, and the container's shipping information would still appear to be legitimate.

Some Megaports Partner Countries Not Fully Participating in the Initiative

Some partner countries are not fully participating in the Megaports Initiative, which raises questions about their long-term commitment to the Initiative and may compromise the security gains expected. For example, Panamanian customs officials—who expressed support for the Initiative— told us that the four Megaports in Panama currently scan less than 1 percent of their container traffic, but terminal operators at two Megaports in Panama stated that they are not willing to conduct additional scanning because they do not believe there are economic benefits to participating in the Initiative. Panamanian customs officials told us that about 85 percent of containers passing through the Panama Canal are transshipped containers. According to the Initiative's program plan, it is difficult to scan transshipped containers because of space and operational constraints. NNSA officials told us that they plan to add a MRDIS unit—a type of mobile radiation detection equipment—to one of the Megaports in Panama to help increase the volume of containers scanned. However, a senior official for the terminal operator at this port told us that they have not yet agreed to the placement of the MRDIS and are concerned about it delaying port operations. Panamanian customs officials told us that a nuclear or radiological incident affecting the Panama Canal would take years to clean up and would cause worldwide economic chaos. Other Megaports partners are also not fully participating in the Megaports Initiative. For example, the Chinese government has agreed to install radiation detection scanning equipment at one port. However, 12 Chinese seaports are ranked in the top 100 highest priority ports, according to NNSA's port prioritization model. NNSA officials told us that they have discussed installing radiation detection equipment at additional seaports in China with Chinese officials, but that future cooperation is uncertain because of the proposed decreases to the Initiative's budget. In addition, China is one of the few countries that does not share scanning data with the Initiative because, according to NNSA officials, the Chinese government believes that trade information contained in the data is proprietary. Oak Ridge National Laboratory analyzes scanning data for Megaports countries on a monthly basis to assess, among other things, whether the monitors are installed and working properly, as well as to collect information on alarm rates and the number of containers scanned. Furthermore, DOE laboratory officials told us that the Chinese government has agreed to scan only exports—not imports, which most other Megaports scan—and required NNSA to purchase Chinese radiation detection equipment instead of U.S. equipment. The Initiative tested the Chinese radiation detection equipment in 2006 with a budget of $180,000 to certify it as reliable for use. In addition, Pakistan—a country of significant nonproliferation concern—has not agreed to install equipment at more than one seaport. NNSA officials told us that they began negotiating in 2006 to expand the program in Pakistan.

Management Weaknesses Impede the Initiative's Effectiveness

We also identified several management weaknesses that reduce the Initiative's effectiveness. Specifically, we have concerns that NNSA (1) has not finalized a long-term plan for ensuring the Initiative's sustainability, (2) does not have sufficient performance measures, and (3) has performed limited testing and deployment of SPM radiation detection equipment. In addition, we found limited coordination between the Megaports Initiative and the Container Security Initiative (CSI), which is implemented by DHS.

NNSA Has Not Finalized a Long-Term Plan for Ensuring the Initiative's Sustainability

NNSA has taken steps to help partner countries prepare to take responsibility for Megaports operations after NNSA transfers the equipment to them. However, NNSA does not have a long-term plan for ensuring the ongoing sustainability of these Megaports operations. For example, NNSA is not systematically following up with partner countries to ensure that they are effectively operating and maintaining Megaportsfunded equipment after NNSA's final transfer of equipment, maintenance, and related financial responsibilities. According to the Initiative's program plan, it is "critical" that the installed systems are appropriately overseen to help counter possible corruption and to ensure that the systems are being used as intended. During a July 2012 meeting with senior officials for the Megaports Initiative, we discussed whether the Initiative had a long-term sustainability plan. After that meeting, NNSA officials provided us with a draft sustainability plan, which addresses how SLD will engage partner countries after the final transfer of radiation detection equipment. For example, the plan states that the Sustainability Program will use a list of metrics to assess how well a country is maintaining its capacity to sustain Megaports operations and that SLD will provide narrowly tailored support to partner countries, if needed, to ensure continued sustainability of operations. For example, SLD might provide technical advice, analyze RPM scanning data, or provide help desk support. According to the Initiative's acting program manager, NNSA plans to finalize this plan in October 2012.

We also found that several impediments could compromise the sustainability of Megaports operations. According to NNSA officials, several countries—including Bangladesh, Djibouti, and Kenya—will likely require ongoing assistance at the end of their 3-year transition periods. Other countries, including Sri Lanka and Colombia, have requested extensions to their 3-year sustainability period, and NNSA officials expect that additional training will be needed to fully transfer Megaports operations to Pakistan. Officials in two countries we visited—the Dominican Republic and Panama—reported concerns about funding the operations and maintenance of Megaports equipment after NNSA transfers responsibility. Furthermore, officials in the five countries we visited reported staffing challenges, including recruiting, retaining, and funding staff who operate Megaports equipment and respond to alarms. For example, CSI officials in one Caribbean country told us that they observed staffing shortages among the partner country's customs officials who work on the Megaports Initiative and that alarms for Megaports RPMs periodically go off for several hours to a week without a response from local customs officials. According to NNSA officials, the partner country has taken steps to resolve these staffing issues by hiring a new lead Megaports officer and new Megaports employees who are to be trained in September 2012. However, a senior official who works for a global terminal operator told us that he is not confident that countries will continue operating the Megaports equipment if the Initiative is eliminated because these countries would not have any diplomatic or economic incentives to continue scanning containers. In fact, this official said that this global terminal operator has already begun looking into the costs of removing radiation scanning equipment in case some governments decide not to continue Megaports operations.

Performance Measures Are Not Sufficient

Performance measures are important to ensure that NNSA can evaluate the Initiative's effectiveness and progress. We have previously reported that agencies successfully assess performance when they use measures that demonstrate results, cover multiple program priorities, and provide useful information for decision making.[23] According to DOE's *Fiscal Year 2011 Annual Performance Report*, NNSA has two performance measures for the Initiative—the total number of foreign seaports where it has installed radiation detection equipment and the cumulative number of Megaports with partner country cost-sharing.

These two performance measures, however, have limitations and do not provide sufficient information for decision making. For example, using the number of foreign seaports where the Initiative has installed radiation detection equipment as a key performance measure does not evaluate the impact and effectiveness of the Initiative. This measure does not, for example, demonstrate whether the equipment is effective, account for the volume of containers being scanned, or indicate whether countries are continuing to use the equipment. As of August 2012, NNSA reported having 42 operational Megaports, but NNSA decommissioned 1 of those Megaports—Southampton in the United Kingdom—in June 2010. In addition, Panamanian customs officials told us that the four Megaports in Panama currently scan less than 1 percent of their container traffic. As such, this performance measure does not provide NNSA decision-makers with adequate information to assess the extent to which the Initiative is fulfilling its mission to deter, detect, and interdict the illicit trafficking of special nuclear and other radioactive materials in the global maritime system.

Furthermore, using the cumulative number of Megaports with country cost-sharing as a measure of the Initiative's performance is not reliable because, as we reported in December 2011, NNSA does not systematically track and verify foreign cost-sharing for the Megaports Initiative, and these estimates are neither validated nor audited.[24] NNSA officials told us that they do not have the authority to audit countries' cost-sharing estimates. In our December 2011 report, we noted that NNSA officials told us that this performance measure was difficult to quantify and that it would be eliminated for fiscal year 2012 as a result of a DOE "performance measures streamlining initiative."

According to a 2011 House Committee on Appropriations report, the committee is concerned that the Initiative's performance measures are not adequate to assess the effectiveness of SLD's activities, and the report states, "the true effectiveness of [radiation] detectors in preventing proliferation is largely dependent on how well individual countries employ these capabilities in their security operations." [25] OMB staff also told us that it is difficult to measure the Megaports Initiative's outcomes because of its limited performance measures and said that the Initiative does not routinely report what materials it has successfully interdicted.

In February 2012, we met with senior NNSA program officials to discuss the status of the Megaports Initiative. These officials told us that the Initiative's current performance measures are insufficient and that they need to develop more effective and useful metrics. For example, NNSA officials told us that one of the primary benefits of the Initiative is deterring terrorists from smuggling nuclear or radiological materials through seaports, but it is difficult to measure the deterrent value of the Initiative.

Limited Testing and Deployment of Radiation Detection Equipment

We identified numerous issues related to NNSA's plans for deploying radiation detection equipment. According to the Initiative's *Alternative Detection Technology Strategy*, SLD acquired 44 SPM units in total—12 purchased by NNSA in 2006 and 32 given to NNSA by DHS in 2010 and 2011.[26] We have reported several times over the last 6 years that SPM technology—also known as advanced spectroscopic portals (ASP)—is not proven.[27] One of the principal findings of our past work reviewing DHS's efforts to develop and procure ASPs was that initial testing was not rigorous enough. Once the testing became more rigorous, ASPs did not perform well enough to warrant development, and the DHS program was subsequently canceled. NNSA officials told us that, after receiving the ASPs from DHS, they changed the name of the equipment to SPMs to avoid the negative connotations associated with the ASP program. NNSA officials also said that they viewed the receipt of the SPMs as an opportunity to test and evaluate this type of technology exclusively for secondary scanning purposes at an actual seaport. However, we visited one foreign seaport where NNSA had deployed an SPM, but the unit was getting limited use as a secondary inspection tool. Local customs officials in this country told us that they only use the SPM about two to three times per day, even though the primary radiation detection equipment triggers about 200 alarms each day.

The Megaports Initiative planned to spend $8.9 million to test and evaluate radiation detection equipment from fiscal years 2006 through 2011, including about $5.2 million for SPM technology. NNSA officials told us that the Initiative does not have a dedicated budget for testing and evaluating radiation detection equipment because the program does not have a technology development mission. However, NNSA also has a program office—the Office of Nonproliferation Research and Development—that, among other things, conducts research, development, and testing of new technologies to improve U.S. capabilities to detect nuclear weapon-usable materials. This office's Proliferation Detection program has an annual budget of more than $200 million, which conducts research and provides technical expertise on next-generation detection capabilities for nuclear materials.

Moreover, the Megaports Initiative has not deployed a significant amount of the equipment for testing, as evidenced by its storing unused radiation detection equipment at a total cost of about $1.3 million each year in both 2010 and 2011.[28] NNSA officials told us that, as of February 2012, NNSA had deployed only 16 of the 44 SPMs to foreign seaports and DOE national laboratories. The agency has not deployed any SPMs to Core Program land border sites. NNSA placed the remaining 28 units in storage along with other types of equipment. In August 2012, the Megaports program manager told us that they were looking into options for removing this equipment from storage by, for example, giving some of the equipment to partner countries or by removing usable components and declaring the rest as surplus inventory.

In addition, in 2008, NNSA purchased a new type of radiation detection equipment—four RDSCs at a total cost of about $18 million—but two of the units are not currently being used. According to NNSA officials, the Initiative purchased these two units with the intention of deploying one to Taiwan and giving one to DHS to address a scanning challenge at a U.S. seaport. However, the Taiwanese port was reconfigured, and the new layout was no longer suitable for using the RDSC. Moreover, DHS decided that it did not want to use the RDSC after conducting an alternatives analysis because DHS did not consider the RDSC to be a viable system suitable for use by U.S. Customs and Border Protection. [29] NNSA considered deploying one of the RDSCs to a port in Italy, but that port lost a major shipping client and decided to reconfigure the port in a way that would eliminate use of straddle carriers. As a result, NNSA is spending an additional $33,000 in 2012 on storage fees for these two idle RDSCs while NNSA officials explore other options for deploying them.

Limited Coordination between Megaports and CSI

The *Megaports Initiative 2011 Program Plan* states that the Initiative has committed to providing radiation detection capability at all CSI ports, which will give CSI additional information in assessing the security risks of maritime containers. (See fig. 6 for a photograph of CSI's nonintrusive imaging system and the Megaports Initiative's radiation detection scanning equipment at a Megaport.) From 2005 through 2008, DOE and DHS signed eight joint declarations of principles with partner countries for eight Megaports to establish coordination on enhancing the security of cargo containers. NNSA officials told us that they provide CSI officials, who are co-located with Megaports, with access to alarm information and that the CSI officials are invited to participate in Megaports training.

However, we found two cases at ports in the Caribbean where the Megaports Initiative and CSI programs were co-located but were not coordinating their efforts to scan containers. DOE and DHS signed joint declarations of principles with the local customs offices in both of these countries. In both countries, CSI officials told us that they use their personal radiation detectors—a type of equipment intended for personal safety but not appropriate for scanning containers—to inspect containers if their primary detection equipment is broken. However, in both countries, the Megaports Initiative had functioning RPMs and handheld radiation detection equipment that, according to the Megaports Initiative's former program manager, CSI officials could have used if they had asked Megaports equipment operators. In addition, CSI officials in one of these countries—who work in a room adjacent to Megaports equipment operators—told us that they periodically observed Megaports alarms going off for several hours to a week without answer. CSI officials told us that they eventually became "annoyed" by the alarms and called the partner country's customs office to report the issue.

Source: GAO.

Figure 6. CSI Nonintrusive Imaging System Next to the Megaports Initiative's Radiation Detection Scanning Equipment.

Moreover, CSI officials said that they do not know if the containers that triggered the alarms were eventually inspected. CSI officials' reluctance to resolve the alarms or contact NNSA officials about this issue raises questions about whether NNSA and DHS are leveraging resources of these two U.S. programs with similar missions.

In an October 2005 report on practices that can help enhance and sustain collaboration among federal agencies, we stated that a focus on results, as envisioned by the Government Performance and Results Act of 1993, implies that federal programs contributing to the same or similar results should collaborate to ensure that goals are consistent and, as appropriate, program efforts are mutually reinforcing. As we reported, agencies can enhance and sustain collaborative efforts by identifying and addressing needs by leveraging resources to support the common outcome.[30] It is unclear to what extent

NNSA and DHS are coordinating or collaborating at other seaports where they are co-located to ensure that their program efforts are mutually reinforcing as envisioned by the Government Performance and Results Act.

Furthermore, NNSA and DHS have entered into joint arrangements with partner countries, but the two agencies do not have written policies or procedures on how to coordinate their activities at ports to best leverage their resources. Without written procedures, the two initiatives may not be taking advantage of an opportunity to more effectively implement their mutual goal of combating nuclear smuggling overseas. In June 2010, we reported that agencies can strengthen their commitment to work collaboratively by articulating their roles and responsibilities in formal documents to facilitate decision making.[31] Such formal documents can include memorandums of understanding, interagency guidance, or interagency planning documents, signed by senior officials in the respective agencies.

CONCLUSION

Under its Megaports Initiative, NNSA has installed radiation detection equipment at more than 40 foreign seaports since 2003 and has trained foreign personnel in partner countries to scan shipping containers entering and leaving these seaports—regardless of the containers' destination—for nuclear or radiological material. However, recent proposed budget cuts have created an uncertain future for the Megaports Initiative. NNSA is working with an interagency work group to evaluate the Megaports Initiative and adjust future plans accordingly. This evaluation and adjustment period provides NNSA with an opportunity to address management weaknesses that we identified during the course of our review and seek to maximize the effectiveness and efficiency of the Initiative whatever its construct in the future.

We found that several Megaports partner countries have not entered into a formal cost-sharing arrangement with NNSA but nonetheless have made contributions to the Initiative. However, NNSA is not systematically tracking information on the amount of funding or other contributions that partner countries informally contribute to the Initiative. We continue to believe that our previous recommendation—that NNSA develop better ways to track and maintain information on foreign cost-sharing—could, if implemented, provide a tangible measure of foreign governments' commitment to NNSA's nuclear nonproliferation programs.

NNSA has not finalized a long-term plan for ensuring the sustainability of Megaports operations after NNSA transfers all equipment maintenance, operations, and related financial responsibilities to partner countries. As we found, several countries have raised concerns about funding the operations and maintenance of Megaports equipment after the United States transfers responsibility. Without a plan in place to ensure the sustainability of Megaports operations in the future, NNSA cannot be confident that this equipment will continue to be used for the purposes intended or if it will be used at all.

NNSA has also not developed performance measures that sufficiently demonstrate the extent to which the Megaports Initiative is effectively fulfilling its mission to deter, detect, and interdict the illicit trafficking of special nuclear and other radioactive materials in the global maritime system. Its current measures of performance—the total number of foreign seaports where NNSA has installed radiation detection equipment and the cumulative number of Megaports with partner country cost-sharing— are not useful or reliable for decision making. Specifically, these measures do not demonstrate whether the equipment is effective, account for the volume of containers being scanned, or indicate whether countries are continuing to use the equipment. As a result, NNSA decision makers and policymakers do not have the necessary information to assess the nonproliferation and deterrent benefits of the Initiative. Without meaningful performance measures, it is difficult to measure the Megaports Initiative's outcomes, which may have contributed to the administration's proposed budget cuts to it.

We found that the Megaports Initiative has planned more than $5 million in spending to test and evaluate SPM radiation detection equipment, which we previously reported is an unproven technology. The Megaports Initiative also does not have a technology development mission; whereas, another NNSA program office does have a mission specifically dedicated to testing and evaluating radiation detection equipment. As a result, we question whether this is an appropriate investment of the Initiative's limited funding, particularly given its uncertain future.

Furthermore, NNSA has not deployed a significant amount of the Megaports Initiative's equipment, as evidenced by the more than $1 million SLD spent in both 2010 and 2011 to store radiation detection scanning equipment—including SPMs and other equipment—that it was unable to install in foreign countries—either through the Megaports Initiative or at Core Program land border sites. As of the time of our review, NNSA officials had not taken steps to remove the remaining 28 SPMs and 2 RDSCs from storage, although they are considering options to do so. Without deploying this equipment for the intended purposes, providing it to other NNSA or DOE offices for research and further testing, or declaring the equipment as surplus inventory, NNSA will continue spending the Initiative's limited resources to store idle equipment, even as the Initiative expects significant budget cuts in fiscal year 2013.

Finally, the *Megaports Initiative 2011 Program Plan* states that the Initiative has committed to providing radiation detection capability at all CSI ports, but we found that NNSA and DHS have engaged in limited coordination at two seaports that we visited where the Megaports Initiative and CSI are co-located. CSI officials in two countries told us that they were using personal radiation detectors—which is a type of equipment intended for personal safety but not appropriate for scanning containers—to scan containers, while more appropriate radiation detection equipment that the Megaports Initiative has provided its partners is located at the same port. It is unclear to what extent NNSA and DHS are coordinating or collaborating at other seaports where they are co-located to ensure that their program efforts are mutually reinforcing as envisioned by the Government Performance and Results Act or how localized the problem is. However, while NNSA and DHS have signed joint arrangements with some partner countries, the two agencies do not have written policies or procedures on how to coordinate at ports to best leverage their resources. Without such written procedures, the Megaports Initiative and CSI may not be taking advantage of an opportunity to more effectively combat nuclear smuggling overseas.

Recommendations for Executive Action

We are making the following five recommendations.

To ensure that the Megaports Initiative is maximizing resources and protecting its investments of $850 million, we recommend that the Administrator of the National Nuclear Security Administration take the following four actions:

- Finalize its draft sustainability plan for ensuring ongoing sustainability of Megaports operations after NNSA transfers all equipment, maintenance, operations, and related financial responsibilities to partner countries.

- Develop and maintain useful and reliable measures to assess the performance of the Megaports Initiative. These measures might include the number and type of interdictions, percentage of global maritime container traffic scanned, or length of time that countries sustain operations of the Megaports equipment.
- Terminate additional expenditures from the Megaports Initiative's budget for testing and evaluating SPMs. Additional testing and evaluation activities, if warranted, should be funded through other NNSA programs with a research and development mission.
- Remove unused Megaports radiation detection equipment—particularly SPMs and RDSCs—from storage on an expedited basis. This could be done, for example, by (1) providing the equipment to DOE national laboratories for research and further testing or (2) declaring the equipment as surplus inventory.

To strengthen efforts to combat nuclear smuggling overseas, we also recommend that the Administrator of the National Nuclear Security Administration and the Secretary of Homeland Security take the following action:

- On a periodic basis, jointly assess the extent to which the Megaports Initiative and CSI are effectively leveraging resources and coordinating at foreign seaports where the two Initiatives are co-located. If the agencies determine based on these joint assessments that there is a need for increased coordination, they should develop written policies or procedures that formally document how to coordinate at ports to best leverage their resources.

AGENCY COMMENTS AND OUR EVALUATION

We provided the Departments of Energy and Homeland Security and the Office of Management and Budget a draft of this report for their review and comment. In written comments, DOE and DHS agreed with our recommendations. DOE and DHS also provided technical comments, which we incorporated as appropriate. On October 3, 2012, we received oral comments from OMB staff, including an OMB branch chief, a program examiner, and staff from the Office of General Counsel. These staff agreed with the facts presented in this report and stated that they concurred with our findings and recommendations. In addition, OMB provided technical comments, which we incorporated as appropriate.

David C. Trimble
Director, Natural Resources and Environment

APPENDIX I: OBJECTIVES, SCOPE, AND METHODOLOGY

The objectives of our review were to assess (1) the status of the Megaports Initiative (Initiative) and the National Nuclear Security Administration's (NNSA) plans for completing and sustaining it and (2) benefits of the Initiative and factors, if any, that reduce its effectiveness. For both objectives, we obtained and analyzed relevant NNSA documentation, including the Megaports Initiative Program Plans for 2011 and 2010, the *2011 Second Line of Defense Implementation Strategy*, and the *2006 Second Line of Defense Strategic Plan*. We also reviewed memorandums of understanding and cost-sharing arrangements between the Department of Energy (DOE) and partner countries, as well as sustainability and joint transition plans. Furthermore, we reviewed and met with NNSA officials responsible for the Initiative's prioritization model, which NNSA developed to help prioritize and select ports for participation in the Initiative. We also analyzed information about the ports NNSA has completed thus far, including information about their cost-sharing, transition dates, and rankings. In addition, we interviewed NNSA officials responsible for implementing and managing the Megaports Initiative about the status of the program, its future plans, sustainability activities, and challenges associated with implementation. We also interviewed NNSA country managers about operational Megaports. Moreover, we interviewed contractors who help implement the Initiative at two DOE national laboratories: (1) Pacific Northwest National Laboratory about equipment, training, and procurement activities and (2) Oak Ridge National Laboratory about its analysis of equipment performance data and alarms. We interviewed officials at the Department of Homeland Security (DHS) and the Department of State about coordination with the Megaports Initiative. Furthermore, we interviewed officials from two global terminal operators—DP World and Hutchinson Port Holdings—regarding their perspectives on the Megaports Initiative. In addition, we visited a nonprobability sample of eight seaports in five countries[1]—Belgium, the Dominican Republic, Jamaica, Panama, and Spain—to observe scanning operations and interview foreign government and port officials.[2] We selected these seaports on the basis of port size, NNSA's priority ranking, NNSA's expenditures at the port, history of cost-sharing with NNSA, length of time in the Megaports Initiative, joint presence with DHS's Container Security Initiative (CSI), and unique characteristics of specific locations. At the seaports we visited, we interviewed foreign government officials using a standard set of questions about their participation in and perspective on the Megaports Initiative. These officials included those from customs offices and offices who handle the response to nuclear emergencies, as well as terminal operators and port authorities. Because we used a nonprobability sample to select the eight Megaports we visited, the observations we made are not generalizeable to other Megaports but serve as examples of challenges and concerns that may affect Megaports operations and sustainability. We also interviewed DHS officials co-located at these seaports and U.S. embassy officials in these countries. To assess the status of the Megaports Initiative and NNSA's plans for completing and sustaining it, we analyzed the Megaports Initiative's budget information and expenditure data for fiscal years 2003 through 2012, as well as projected costs through fiscal year 2018. We interviewed knowledgeable NNSA officials to assess the reliability of the data and discussed such issues as data entry, access, quality control procedures, and the accuracy and completeness of the data. We determined that the data provided to us were of sufficient reliability for the purposes of this review. Moreover, we reviewed NNSA's budget request and related budget documents for fiscal year 2013 and spoke with Office of Management and Budget staff about proposed changes to the Megaports Initiative's fiscal year 2013 budget. We also requested and analyzed data on Megaports partner countries' cost-sharing contributions. Because these contributions are a combination of NNSA and foreign country estimates, we did not have a way to verify these data. As such, these data are of undetermined reliability. We are nonetheless reporting these cost-sharing data to provide some context on cost-sharing contributions made by partner countries. To assess the benefits of the Megaports Initiative and any factors that reduce its effectiveness, we obtained and reviewed lessons learned that NNSA officials document in a central database, performance measures in DOE's fiscal year *2011 Annual Performance Report*, plans for radiation detection equipment outlined in the Megaports Initiative's *2012 Alternative Detection Technology Strategy*, and information on the Initiative's costs for storing radiation detection equipment. We also reviewed DHS's U.S. Customs and Border Protection's directive on its domestic radiation detection program. In addition to the interviews listed above, we also interviewed NNSA headquarters, Pacific Northwest National Laboratory, and foreign officials about the benefits and challenges they have encountered in implementing the Initiative. We also conducted an interview with officials from DOE's Office of Intelligence and Counterintelligence regarding the Megaports Initiative's process for selecting ports. We conducted this performance audit from August 2011 to October 2012 in accordance with generally accepted government auditing standards. Those standards require that we plan and perform the audit to obtain sufficient, appropriate evidence to provide a reasonable basis for our findings and conclusions based on our audit objectives. We believe that the evidence obtained provides a reasonable basis for our findings and conclusions based on our audit objectives.

APPENDIX II. RADIATION DETECTION EQUIPMENT USED BY THE MEGAPORTS INITIATIVE

Equipment type	Description and uses	Limitations	Number of units deployed to Megaports	Number of units in storage
Radiaton Detection Portal Monitor (RPM)				
	• RPMs scan shipping containers on trucks or railcars as they are entering or exiting seaports or as shipping containers are being off-loaded from sea vessels. • Used for primary scanning purposes.	• RPMs cannot distinguish the type of material causing an alarm.	458 vehicle RPMs and 58 rail RPMs	134 vehicle RPMs and 32 rail RPMs
Mobile Radiaton Detection Identification System (MRDIS)				
	• Used to scan containers at high-volume transshipment ports. • After the MRDIS is parked at a scanning location, trucks drive containers through it in the same manner as a fixed RPM. • Can be driven to different locations in a port. • Used for primary scanning purposes, but also sometimes equipped with technology that car conduct secondary inspections.	• MRDISs with only primary scanning capabilities cannot distinguish the type of material causing an alarm.	13	0
Radiaton Detection Straddle Carrier (RDSC)				
	• Used to scan containers at high-volume transshipment ports that use straddle carriers for port operations. • Can be driven through rows of containers stacked one to three high. • Equipped with technology for both primary and secondary scanning.	• RDSCs are designed to operate in ports where containers are stacked up to three high and that currently use straddle carriers to move containers, but not all ports have such a configuration.	2	2

Equipment type	Description and uses	Limitations	Number of units deployed to Megaports	Number of units in storage
Spectroscopic Portal Monitor (SPM)[a]				
	• SPMs look similar to standard RPMs but have an additional capability to identify the source of radiation. • The Initiative has deployed some SPMs in select locations. • Used for secondary inspections.	• NNSA is currently using the SPM units at select seaports for testing and evaluation purposes	17[b]	21
Radioisotope Identification Device (RIID)				
	• RIIDs are handheld devices used to pinpoint the source of an alarm in a shipping container and determine the type of radioactive material present. • Used for secondary scanning purposes.	• GAO has previously reported that completing a secondary inspection with a RIID typically takes several minutes.[c]	334	31
Personal Radiation Detector (PRD)				
	• PRDs, or pagers, are small detectors that can be worn on an individual's belt to continuously monitor radiation levels in the immediate area of the officials. • PRDs help to ensure personal safety.	• PRDs are not appropriate for scanning shipping containers.	958	290

Source: GAO analysis of NNSA documents; GAO (photographs).

[a] SPMs are formerly known as advanced spectroscopic portal (ASP) monitors.

[b] NNSA has also deployed four SPM units to DOE's national laboratories for testing and evaluation purposes.

[c] GAO, *Combating Nuclear Smuggling: DHS's Cost-Benefit Analysis to Support the Purchase of New Radiation Detection Portal Monitors Was Not Based on Available Performance Data and Did Not Fully Evaluate All the Monitors' Costs and Benefits*, GAO-07-133R (Washington, D.C.: Oct. 17, 2007).

Appendix III: Information about NNSA Megaports

Country	Megaport	Priority ranking, by tier[a]	Size[b]	Operational date[c]	Transition date[d]	CSI port	Percentage of totalcontainer volumescanned (estimated)
Argentina	Buenos Aires	50-100	Medium	Sept. 2012[e]	Sept. 2015	Yes	97[f]
Bahamas	Freeport	50-100	Small	June 2006	Sept. 2014[g]	Yes	85
Bangladesh	Chittagong	1-50	Medium	June 2011	July 2014	No	81
Belgium	Antwerp	1-50	Large	April 2007	Jan. 2010	Yes	73
	Zeebrugge	100-150	Medium	April 2009	Sept. 2010	Yes	75
Cambodia	Sihanoukville	50-100	Small	Sept. 2011	Sept. 2014	No	100
China	Shanghai	1-50	Medium	Dec. 2011	Dec. 2014	Yes	81
Colombia	Cartagena	100-150	Small	Sept. 2008	Apr. 2012[h]	Yes	33
Djibouti	Djibouti	50-100	Small	March 2011	March 2014	No	65
Dominican Republic	Caucedo	100-150	Small	Sept. 2008	Sept. 2011	Yes	99
Greece	Piraeus[i]	150-200	Medium	Aug. 2004	Oct. 2007	Yes	80
Honduras	Puerto Cortes	100-150	Small	April 2007	Sept. 2012[j]	Yes	100
Israel	Ashdod	1-50	Medium	Nov. 2010	Nov. 2013	Yes	95
	Haifa	1-50	Small	Jan. 2008	Dec. 2015[k]	Yes	31
Jamaica	Kingston	50-100	Small	June 2009	March 2015[l]	Yes	85
Jordan	Aqaba	50-100	Small	Sept. 2010	Sept. 2013[m]	No	100
Kenya	Mombasa	50-100	Medium	Feb. 2011	Feb. 2014	No	65
Lebanon	Beirut	1-50	Small	Aug. 2010	July 2013	No	47
Malaysia	Klang	1-50	Large	Sept. 2009	Sept. 2013	Yes	70
	Tanjung Pelepas	1-50	Medium	Sept. 2010	Sept. 2013	Yes	90
Malta	Marsaxlokk	50-100	Small	June 2012	June 2015	No	49
Mexico	Altamira[i]	150-200	Small	Dec. 2011	May 2013	No	82
	Lazaro Cardenas[i]	150-200	Small	June 2009	May 2013	No	50

Country	Megaport	Priority ranking, by tiera	Sizeb	Operational datec	Transition dated	CSI port	Percentage of totalcontainer volumescanned (estimated)
	Manzanillo	100-150	Small	Nov. 2009	May 2013	No	76
	Veracruz	100-150	Small	March 2009	May 2013	No	85
Netherlands	Rotterdam	1-50	Small	April 2004	Oct. 2007	Yes	70
Oman	Salalah	100-150	Medium	Sept. 2008	Oct. 2015[n]	Yes	36
Pakistan	Qasim	50-100	Small	Oct. 2007	TBD[o]	Yes	10
Panama	Balboa	100-150	Small	Sept. 2008	Dec. 2011	Yes	20[p]
	Colon Container Terminal[i]	150-200	Small	Jan. 2010	Jan. 2013	Yes	5[p]
	Cristobal[i]	150-200	Small	Jan. 2010	Jan. 2013	No	5[p]
	Manzanillo International Terminal[i]	150-200	Small	Sept. 2008	Dec. 2011	Yes	10[p]
Philippines	Manila	1-50	Medium	Feb. 2007	Sept. 2014	No	100
	Port of Cebu	50-100	Small	Sept. 2012[e]	Sept. 2014	No	99[f]
Portugal	Lisbon[i]	150-200	Medium	Sept. 2009	June 2013[q]	Yes	90
Singapore	Singapore	1-50	Small	April 2006	June 2012	Yes	14
South Korea	Busan	1-50	Small	March 2009	March 2014[r]	Yes	28
Spain	Algeciras[i]	150-200	Medium	March 2006	July 2014[s]	Yes	9
	Barcelona	100-150	Medium	June 2011	July 2014	Yes	55
	Valencia	50-100	Medium	Sept. 2010	Apr. 2014	Yes	49
Sri Lanka	Colombo	1-50	Medium	Jan. 2006	Apr. 2013[t]	Yes	31
Taiwan	Kaohsiung	1-50	Large	Sept. 2009	Sept. 2012	Yes	98
Thailand	Laem Chabang	1-50	Large	May 2007	Apr. 2014[u]	Yes	96
United Kingdom	Southampton[v]	1-50	Small	Oct. 2007	June 2010	Yes	0
Vietnam	Cai Mep[i]	150-200	Medium	Sept. 2012[e]	Sept. 2015	No	95[f]

Source: GAO analysis of NNSA documents.

[a]The priority ranking listed is NNSA's prioritization ranking by countries' adjusted score. NNSA prioritizes countries with a raw score and then adjusts the rankings by removing countries from the list that (1) have low container volume, (2) NNSA cannot work with for diplomatic or policy reasons, or (3) NNSA or other organizations conduct similar work at that port. For security purposes, we have clustered the rankings into tiers: 1-50, 50-100, 100-150, and 150-200.

[b] NNSA estimates the size of the Megaports installation—small, medium, or large—based on the number of RPM units needed to scan as many shipping containers as possible. This may be different from the physical size of the port.

[c] The operational date is the date that the partner country begins operating the Megaports radiation detection scanning equipment.

[d] The transition date is the date that NNSA currently plans to transfer all equipment, maintenance, operations, and related financial responsibilities to partner countries.

[e] These dates are NNSA's current estimates.

[f] These ports were constructed during fiscal year 2012 and, therefore, the percentages of containers scanned are NNSA's expected scanning volumes, rather than actual estimated figures.

[g] From 2006 through 2010, NNSA tested a prototype of the RDSC in Freeport. NNSA began operating a second generation RDSC in June 2010. NNSA had discussions with the Bahamian government in June 2012 and intends to transfer responsibility for the RDSC in September 2014.

[h] NNSA officials told us that Colombian Customs requested an extension of the transition date in June 2010.

[I] According to NNSA documents, these countries were added to the Megaports prioritization list because they were (1) already operational prior to NNSA developing its prioritization model or (2) a specific port of interest for the Megaports Initiative.

[j] According to NNSA officials, the sustainability period was extended from 2009 through 2012 because of political turmoil in Honduras.

[k] The Megaports Initiative began a pilot project at the Port of Haifa in January 2008, which it dismantled in January 2012. NNSA expects to complete full installation at Haifa in December 2012 and plans to transition the equipment in December 2015.

[l] The Megaports Initiative has installed radiation detection equipment at two terminals in Kingston. The first terminal began operations in June 2009, and NNSA transitioned the equipment to Jamaican Customs in June 2012. The second terminal began operations in March 2012, and NNSA expects to transition the equipment in March 2015.

[m] The Megaports Initiative installed equipment at four different locations in Jordan—including at one port in Aqaba and at three land border sites—and will provide each location with 3 years of maintenance, as follows: Aqaba, September 2010 through September

2013; Jaber crossing, March 2011 through March 2014; Al Karama crossing, June 2011 through June 2014; and Al Omari crossing, May 2012 through May 2015. NNSA considers all work in Jordan as one Megaport.

[n] NNSA installed radiation detection equipment at the Port of Salalah, Oman, in September 2008 but has since updated some of the MRDIS radiation detection equipment with second generation units, which NNSA expects to be operational by October 2012. NNSA intends to transfer the equipment to the government of Oman in October 2015.

[o] The Port of Qasim is the only port remaining under the Secure Freight Initiative and, according to NNSA, there is no planned transition date at this time. The Secure Freight Initiative is a DHS and DOE program at selected ports that aims to scan 100 percent of U.S.-bound container cargo for nuclear and radiological materials overseas using non-intrusive imaging systems and radiation detection equipment.

[p] These scanning estimates are according to the *Megaports Initiative 2011 Program Plan*. However, Panamanian customs officials that we met with told us that they currently scan less than 1 percent of their container traffic at their four Megaports.

[q] NNSA installed Megaports equipment at the Port of Lisbon, Portugal, in two phases. The first phase began operating in September 2009, and the second began operating in June 2010. NNSA plans to transfer the equipment to the Portuguese government in June 2013.

[r] NNSA and DHS started the Secure Freight Initiative at the Port of Busan in March 2009, and DHS decommissioned its equipment in March 2010. The Megaports Initiative signed a new Memorandum of Understanding in September 2010 and restarted the radiation detection system in March 2011.

[s] NNSA transitioned responsibility for RPM units at Algeciras to Spanish Customs in December 2010 and plans to transition responsibility for a SPM unit in July 2014.

[t] NNSA added a SPM unit in fiscal year 2009, and in fiscal year 2010 NNSA began an upgrade project to improve Megaports operations at Colombo. NNSA aims to transition responsibility of the equipment to the Sri Lankan government in April 2013.

[u] NNSA and the Thai government negotiated a longer transition period because of initial system issues and a request from the Thai government to have a 5-year transition period.

[v] NNSA began operations in Southampton in October 2007 through the Secure Freight Initiative, but DHS concluded Secure Freight Initiative operations in June 2010. NNSA terminated operations of the radiation detection equipment in June 2010 and gave an SPM unit to the United Kingdom Border Agency, which relocated the equipment to another port.

Appendix IV: List of Operational Megaports and Megaports Being Implemented

Country	Megaport
Argentina	Buenos Aires[a]
Bahamas	Freeport
Bangladesh	Chittagong
Belgium	Antwerp
	Zeebrugge
Cambodia	Sihanoukville
China	Shanghai
Colombia	Cartagena
Djibouti	Djibouti
Dominican Republic	Caucedo
Greece	Piraeus
Honduras	Puerto Cortes
Israel	Ashdod
	Haifa
Jamaica	Kingston
Jordan	Aqaba
Kenya	Mombasa
Lebanon	Beirut
Malaysia	Klang
	Tanjung Pelepas
Malta	Marsaxlokk
Mexico	Altamira
	Lazaro Cardenas
	Manzanillo
	Veracruz
Netherlands	Rotterdam
Oman	Salalah
Pakistan	Qasim
Panama	Balboa
	Colon Container Terminal
	Cristobal
	Manzanillo International Terminal
Philippines	Manila
	Port of Cebu[a]
Portugal	Lisbon
Singapore	Singapore
South Korea	Busan
Spain	Algeciras
	Barcelona
	Valencia
Sri Lanka	Colombo
Taiwan	Kaohsiung
Thailand	Laem Chabang
United Kingdom	Southampton
Vietnam	Cai Mep[a]

[a] NNSA expects these Megaports to become operational in September 2012.

APPENDIX V: MEGAPORTS INITIATIVE'S TOTAL BUDGET AND EXPENDITURES, FISCAL YEARS 2003 THROUGH 2012

Dollars in thousands						
Fiscal year	Enacted budget (current dollars)	Supplemental funding (current dollars)	Enacted budget (constant dollars)	Supplemental funding (constant dollars)	Total budget (constant dollars)	Total Expenditures (constant dollars)
2003	$15,000	$84,000	$18,340	$102,702	$121,042	$1,341
2004	13,000	0	15,501	0	15,501	56,381
2005	15,000	29,000	17,323	33,492	50,815	60,921
2006	73,929	0	82,575	0	82,575	57,065
2007	40,118	71,000	43, 522	77,025	120,547	88,673
2008	132,047	0	140,003	0	140,003	102,749
2009	108,091	0	112,973	0	112,973	136,397
2010	174,264	0	180,461	0	180,461	167,344
2011	124,884	0	126,824	0	126,824	145,126
2012	132,670	0	132,670	0	132,670	33,787[a]
Total	**$829,003**	**$184,000**	**$870,193**	**$213,219**	**$1,083,412**	**$849,784**

Source: GAO analysis of NNSA data.

[a] Fiscal year 2012 expenditures are as of December 2011.

End Notes

[1] Weapon-usable nuclear materials are highly enriched uranium, uranium-233, and any plutonium containing less than 80 percent of the isotope plutonium-238. Such materials are often referred to as strategic special nuclear materials.

[2] Twenty-foot shipping containers, referred to as a twenty-foot equivalent unit, are the industry standard, and now cargo volume and vessel capacity are commonly measured in twenty-foot equivalent units.

[3] NNSA was created by the National Defense Authorization Act for Fiscal Year 2000, Pub. L. No. 106-65 (1999). It is a separate semi-autonomous agency within DOE, with responsibility for the nation's nuclear weapons, nonproliferation, and naval reactors programs.

[4] Scannable shipping volume refers to the amount of inbound, outbound, and transshipped containers that NNSA estimates can be scanned using existing radiation detection equipment. Transshipped containers are containers that are unloaded from one ship to a seaport for a short period of time before being loaded onto another ship.

[5] GAO, *Preventing Nuclear Smuggling: DOE Has Made Limited Progress in Installing Radiation Detection Equipment at Highest Priority Foreign Seaports*, GAO-05-375 (Washington, D.C.: Mar. 31, 2005).

[6] These seaports included in Belgium: Antwerp and Zeebrugge; in Dominican Republic: Caucedo; in Jamaica: Kingston; in Panama: Balboa, Colon Container Terminal, and Manzanillo International Terminal; and in Spain: Barcelona.

[7] For our 2005 report, we visited Megaports in Rotterdam, the Netherlands, and Piraeus, Greece.

[8] CSI targets and examines high-risk containers for weapons of mass destruction before they are shipped to the United States.

[9] DOE manages the largest laboratory system of its kind in the world. The mission of DOE's 22 laboratories has evolved. Originally created to design and build atomic weapons, these laboratories have since expanded to conduct research in many disciplines—from high-energy physics to advanced computing.

[10] NNSA had spent about $850 million on Megaports Initiative activities as of December 2011. Unless otherwise indicated, all dollars are in 2012 constant dollars.

[11] Scannable volume refers to the estimated amount of inbound, outbound, and transshipped cargo that can be scanned utilizing existing Megaports Initiative equipment.

[12] GAO-05-375.

[13] These data are of undetermined reliability. Because these are NNSA and foreign country estimates, we did not have a way to verify this information.

[14] GAO, *Nuclear Nonproliferation: Action Needed to Address NNSA's Program Management and Coordination Challenges*. GAO-12-71 (Washington, D.C.: Dec. 14, 2011).

[15] GAO, *Nuclear Nonproliferation: DOE's International Radiological Threat Reduction Program Needs to Focus Future Efforts on Securing the Highest Priority Radiological Sources*, GAO-07-282 (Washington, D.C.: Jan. 31, 2007).

[16] NNSA's projected costs are in current dollars.

[17] The other $73.0 million would be allocated to the Core Program. As discussed above, Core Program activities include funding the installation of radiation detection equipment at land borders, airports, and strategic seaports in Russia, other former Soviet Union countries, and Eastern Europe.

[18] As of October 2012, Congress had passed a continuing resolution generally funding agencies at fiscal year 2012 levels through March of 2013 but had passed no full-year appropriation for DOE reflecting the proposed budget cuts. The House Committee on Appropriations, in the report accompanying H.R. 5325, an Energy and Water Development Appropriations Bill that passed the House of Representatives in April, recommended enactment of the administration-proposed budget cuts for the Megaports Initiative. The Senate Appropriations Committee, in the report accompanying S. 2465, an Energy and Water Development Appropriations Bill that has not passed the Senate, expressed concern over the proposed cuts to SLD activities, including the Megaports Initiative, and recommended an increase in funding of $57 million for International Nuclear Materials Protection and Cooperation activities as a whole. Specifically, the Senate committee expressed the view that the budget level proposed for the Megaports Initiative would not be sufficient to sustain already deployed systems, retain expert personnel, and meet international obligations to deploy additional radiation detection equipment.

[19] OMB works with the President of the United States and executive branch departments and agencies to, among other things, manage the administration's budget development each fiscal year.

[20] Masking is the use of naturally occurring radioactive material to make the radiation emitted by smuggled material appear to be caused by innocent cargo. In contrast, shielding blocks radiation from being emitted.

[21] Medical isotopes are used to treat cancer or conduct medical diagnoses.

[22] Natural sources of radiation, which are usually harmless, occur in a wide variety of common items and consumer goods, such as fertilizer, bananas, and ceramic tiles.

[23] GAO, *Executive Guide: Effectively Implementing the Government Performance and Results Act*, GAO/GGD-96-118 (Washington, D.C.: June 1, 1996).

[24] GAO-12-71.

[25] H.R. Rep. No. 112-118 at 137 (accompanying H.R. 2354, Energy and Water Development Appropriations Bill, 2012).

[26] According to NNSA, in 2006 the average cost of an SPM unit was about $600,000.

[27] In prior reports, we have questioned DHS's plans to deploy spectroscopic technology to scan cargo at ports of entry because (1) DHS's cost analysis of the ASP program was not sound and (2) DHS did not fully disclose the limitations of the technology's test results. In February 2010, DHS announced that it was scaling back its plan for development and use of this technology. See, for example: GAO, *Combating Nuclear Smuggling: Lessons Learned from DHS Testing of Advanced Radiation Detection Portal Monitors*, GAO-09-804T (Washington, D.C.: June 25, 2009); GAO, *Combating Nuclear Smuggling: DHS's Phase 3 Test Report on Advanced Portal Monitors Does Not Fully Disclose the Limitations of the Test Results*, GAO-08-979 (Washington, D.C.: Sept. 30, 2008); and GAO, *Combating Nuclear Smuggling: DHS has Developed Plans for Its Global Nuclear Detection Architecture, but Challenges Remain in Deploying Equipment*, GAO-12-941T (Washington, D.C., July 26, 2012).

[28] According to the Initiative's former program manager, the Megaports Initiative and the Core Program each pay half of these costs.

[29] According to DHS officials, the department was undergoing an alternatives analysis for the on-dock rail program, in which the RDSC was one of the alternatives. DHS decided not to take the RDSC until the alternatives analysis was concluded and only if the RDSC was determined to be a viable alternative. Through its analysis, DHS determined there were cost and sustainability issues associated with the RDSC—including high acquisition, operations, and maintenance costs, as well as costs associated with paying RDSC drivers. DHS officials also said that U.S. Customs and Border Protection has since developed cost-sharing initiatives with its stakeholders, which may allow the RDSC to become a viable scanning solution.

[30] GAO, *Results-Oriented Government: Practices That Can Help Enhance and Sustain Collaboration among Federal Agencies*, GAO-06-15 (Washington, D.C.: Oct. 21, 2005).

[31] GAO, *National Security: Key Challenges and Solutions to Strengthen Interagency Collaboration*, GAO-10-822T (Washington, D.C.: June 9, 2010) and GAO, *Live Animal Imports: Agencies Need Better Collaboration to Reduce the Risk of Animal-Related Diseases*, GAO-11-9 (Washington, D.C.: Nov. 8, 2010).

End Notes for Appendix I

[1] These seaports included: Belgium: Antwerp and Zeebrugge; Dominican Republic: Caucedo; Jamaica: Kingston; Panama: Balboa, Colon Container Terminal, and Manzanillo International Terminal; and Spain: Barcelona.

[2] or our 2005 report, we visited Megaports in Rotterdam, the Netherlands, and Piraeus, Greece.

In: Prevention of Nuclear Smuggling … ISBN: 978-1-62417-907-5
Editor: Valere Pruitt

Chapter 2

COMBATING TERRORISM: ADDITIONAL STEPS NEEDED TO ENHANCE FOREIGN PARTNERS' CAPACITY TO PREVENT TERRORIST TRAVEL*

United States Government Accountability Office

WHY GAO DID THIS STUDY

Eliminating the threat of terrorist attacks continues to be a primary U.S. national security focus. According to the 9/11 Commission, constraining the mobility of terrorists is one of the most effective weapons in fighting terrorism. This report (1) describes key gaps the U.S. government has identified in foreign countries' capacity to prevent terrorist travel overseas, (2) evaluates how U.S. capacity-building efforts address those gaps, and (3) assesses the extent to which the U.S. government is measuring progress in its efforts to close those gaps. To identify the key gaps, GAO reviewed governmentwide assessments of vulnerabilities in the international travel system. GAO reviewed the strategies and documentation of U.S. agencies funding and/or implementing foreign capacity-building efforts to prevent

* This is an edited, reformatted and augmented version of United States Government Accountability Office, Publication No. GAO-11-637, dated June 2011.

terrorist travel overseas, including those of the Departments of State (State)—which coordinates U.S. efforts overseas— Defense (DOD), Homeland Security (DHS), Justice (DOJ), and the U.S. Agency for International Development (USAID). GAO also interviewed officials from the National Security Staff, of the National Security Council (NSC), which oversees counterterrorism policy. GAO met with these agencies and conducted field work in Kenya, Pakistan, the Philippines, and Thailand.

WHAT GAO RECOMMENDS

GAO recommends that (1) State develop a mechanism to improve coordination of various agencies' efforts to provide fraudulent travel document training to foreign partners, and (2) NSC develop a mechanism to measure, track, and report on overall progress toward the goal of enhancing foreign partners' capacity to prevent terrorist travel overseas. State concurred with the first recommendation. NSC did not comment on the draft report.

WHAT GAO FOUND

The U.S. government has identified four key gaps in foreign countries' capacity to prevent terrorist travel overseas, as shown below:

Key Gaps in Foreign Countries' Capacity to Prevent Terrorist Travel Overseas

Key gaps	Illustrative examples
1.Sharing information about known and suspected terrorists	Lack of a database system with terrorist screening information (identifying or biographical information on people with known or suspected links to terrorism)
2.Addressing the use of fraudulent travel documents	Manufacture and use of fraudulent travel documents
3.Ensuring passport issuance security	Easily counterfeited or doctored low-quality passports
4.Combating corruption in passport issuance and immigration agencies	Corrupt immigration officials that allow terrorists to pass through checkpoints

Source: GAO analysis of National Counterterrorism Center and Human Smuggling and Trafficking Center documents.

U.S. government foreign capacity-building programs and activities address these gaps to varying degrees. For instance, as one of the U.S. efforts to enhance foreign partners' sharing of information about known and suspected terrorists, State's Terrorist Interdiction Program provides participating countries with hardware and software to develop, maintain, and use terrorist screening information. In fiscal year 2010, nearly 150 ports of entry overseas were using this program. With regard to addressing the use of fraudulent travel documents, GAO found the potential for overlap and duplication since seven components of three federal agencies are involved in providing training on fraudulent travel document recognition to foreign government officials, with no mechanism to coordinate such training. In two countries GAO visited, there was a lack of collaboration among agencies funding and implementing training on this topic. For example, in Pakistan, State and DHS were both planning to hold fraudulent travel document training for the same Pakistani agency during the same month without knowing of the other's plans. Regarding helping countries improve the security of their passport issuance, State and USAID have multiple efforts, including State's Bureau of Consular Affairs bringing delegations from foreign passport offices to the United States for briefings at passport-related agencies. Finally, the U.S. government has many efforts aimed at combating corruption overseas, such as encouraging countries to pass anticorruption laws. While these efforts are not aimed specifically at countries' passport and immigration agencies, they are intended to improve the effectiveness of all government functions.

The U.S. government lacks performance measures to assess governmentwide progress in closing the key gaps in foreign partners' capacity to prevent terrorist travel overseas. None of the governmentwide or individual agency strategic documents GAO reviewed contained such measures. While components of State and DOJ have some performance measures related to information sharing, these measures do not provide decision makers with comprehensive information on governmentwide progress in enhancing foreign partners' capacity.

ABBREVIATIONS

ATA	Anti-Terrorism Assistance
CA	Bureau of Consular Affairs
CBP	U.S. Customs and Border Protection
COCOM	Combatant Command

DHS	Department of Homeland Security
DOD	Department of Defense
DOJ	Department of Justice
DS	Bureau of Diplomatic Security
DSCA	Defense Security Cooperation Agency
FBI	Federal Bureau of Investigation
HSPD-6	Homeland Security Presidential Directive 6
HSTC	Human Smuggling and Trafficking Center
ICAO	International Civil Aviation Organization
ICE	U.S. Immigration and Customs Enforcement
ICITAP	International Criminal Investigative Training Assistance Program
ILEA	International Law Enforcement Academy
INL	Bureau of International Narcotics and Law Enforcement Affairs
INTERPOL	International Criminal Police Organization
IO	Bureau of International Organization Affairs
ISN	Bureau of International Security and Nonproliferation
MCC	Millennium Challenge Corporation
NCTC	National Counterterrorism Center
PISCES	Personal Identification, Secure Comparison, and Evaluation System
PM	Bureau of Political-Military Affairs
S/CT	Office of the Coordinator for Counterterrorism
State	Department of State
TIP	Terrorist Interdiction Program
TSA	Transportation Security Administration
USAID	U.S. Agency for International Development
USCG	United States Coast Guard

June 30, 2011

The Honorable Susan M. Collins
Ranking Member
Committee on Homeland Security
and Governmental Affairs
United State Senate

The Honorable John Tierney
Ranking Member
Subcommittee on National Security,
Homeland Defense, and Foreign Operations Committee
on Oversight and Government Reform
House of Representatives

Eliminating the threat of terrorist attacks continues to be a primary U.S. national security focus. According to the 9/11 Commission, constraining the mobility of terrorists overseas is one of the most effective weapons in fighting terrorism and constraining terrorist travel internationally should become a vital part of counterterrorism strategy. In 2006, the National Counterterrorism Center (NCTC)[1] released the *National Strategy to Combat Terrorist Travel,* which established governmentwide goals for preventing terrorist travel and identified enhancing the capacity of partner nations as one of two pillars supporting that strategy. The attempted attack on a Detroit-bound airliner on December 25, 2009, involving an individual allegedly affiliated with Al Qaeda in the Arabian Peninsula who had traveled from Nigeria and transited through Amsterdam, highlights the continued critical importance of U.S. foreign partners' capacity to stop terrorists before they travel to the United States.

This report evaluates the U.S. government's efforts to close gaps identified in the capacity of foreign partners to stop terrorists from traveling across international borders. Specifically, this report (1) describes the key gaps the U.S. government has identified in foreign countries' capacity to prevent terrorist travel overseas, (2) evaluates how U.S. foreign capacity-building efforts address those gaps, and (3) assesses the extent to which the U.S. government is measuring progress in its efforts to close those gaps.

To address these objectives, we reviewed the *National Strategy to Combat Terrorist Travel* and U.S. government assessments of vulnerabilities in the international travel system that could be exploited by terrorists. We also reviewed documentation on U.S. capacity-building efforts to help countries prevent terrorist travel overseas, from the Departments of State (State), Defense (DOD), Homeland Security (DHS), and Justice (DOJ); the U.S. Agency for International Development (USAID); and the NCTC. For the purposes of this engagement, we define terrorist travel as movements of known or suspected terrorists overseas, crossing international borders outside of the United States by land, sea, or air. We also interviewed knowledgeable officials in Washington D.C., including those from the White House National Security Staff and relevant components of the intelligence community.

To obtain examples of U.S. efforts and more in-depth understanding of specific programs, we conducted fieldwork in four countries selected using criteria that included participation in U.S. capacity-building programs designed to prevent terrorist travel, designation as a terrorist safe haven, presence of key U.S. agency personnel at post, or coverage of key regions to counterterrorism. Based on these criteria, we traveled to Kenya, Pakistan, the Philippines, and Thailand.[2] In each location, we met with U.S. government personnel involved in capacity building to prevent terrorist travel abroad to learn about the types of activities they undertake, how they measure progress, and their reported results. We also met with foreign government officials to learn about the challenges they face in improving their ability to prevent terrorist travel abroad and their perspectives on the effectiveness of U.S. efforts.

We conducted this performance audit from July 2010 to June 2011 in accordance with generally accepted government auditing standards. Those standards require that we plan and perform the audit to obtain sufficient, appropriate evidence to provide a reasonable basis for our findings and conclusions based on our audit objectives. We believe that the evidence obtained provides a reasonable basis for our findings and conclusions based on our audit objectives. Appendix I provides a more detailed description of our scope and methodology.

BACKGROUND

The vulnerability of the international travel system to terrorists crossing international borders to perpetrate terrorist acts against countries' citizens became a major concern after the terrorist attacks of September 11, 2001. Subsequently, Congress passed a series of laws that called for various measures to address weaknesses in U.S. and other countries' foreign travel systems. The Intelligence Reform and Terrorism Prevention Act of 2004 directed the NCTC to submit to Congress a strategy for combating terrorist travel.[3] In 2006, the NCTC issued the *National Strategy to Combat Terrorist Travel.* One of the strategy's two pillars was to enhance U.S. and foreign partner capabilities to constrain terrorist mobility overseas. Among the pillar's objectives were to suppress terrorists' ability to cross international borders and help partner nations build capacity to limit terrorist travel.

The Intelligence Reform and Terrorism Prevention Act of 2004 established the interagency Human Smuggling and Trafficking Center (HSTC) to serve, in part, as a clearinghouse for all U.S. agency information on preventing terrorist travel, and to submit annual assessments of vulnerabilities in the foreign travel system that may be exploited by international terrorists. Later, the Implementing Recommendations of the 9/11 Commission Act of 2007 called for the HSTC to serve as the focal point for interagency efforts to integrate and disseminate intelligence and information related to terrorist travel.[4] The 2007 Act directed DHS, with the cooperation of other relevant agencies, to ensure that HSTC have no less than 40 full-time positions, including, as appropriate, detailees from DHS, State, DOJ, DOD, NCTC, the Central Intelligence Agency, the National Security Agency, and the Department of the Treasury.[5] Presently, DHS' U.S. Immigration and Customs Enforcement (ICE) provides the director of the center, which includes personnel from State, DHS, and the U.S. intelligence community.

NCTC and HSTC jointly issued the first terrorist travel vulnerability assessment in 2005, and HSTC issued additional terrorist travel vulnerability assessments in 2008 and 2009. The assessments synthesize information and analyses from key stakeholders throughout the U.S. government. Specifically, HSTC officials review intelligence and other information from all relevant agencies; attend interagency working groups, interagency intelligence meetings, and other coordination meetings related to terrorist travel; review open source information from banks, nongovernmental organizations, and multinational organizations; and consult with agencies responsible for implementing programs. All relevant agencies are given the opportunity to review and comment on the drafts.

Various U.S. agencies and subagencies are involved in providing capacity building related to enhancing countries' ability to prevent terrorist travel abroad. As shown in figure 1, counterterrorism as a whole, including preventing terrorist travel, is overseen at the policy level by the Office of the Director of National Intelligence and by the National Security Council. The Director of NCTC reports both to the President regarding executive branch-wide counterterrorism planning, and to the Director of National Intelligence regarding intelligence matters. NCTC follows the policy direction of the President and the National Security Council. State, DHS, DOD, and DOJ fund and/or implement the majority of the capacity-building programs. Within the Department of State, the Office of the Coordinator for Counterterrorism (S/CT), in addition to funding and implementing capacity-building programs, has a leading role in developing coordinated strategies to defeat terrorists abroad and securing the cooperation of international partners. S/CT works with all appropriate elements of the U.S. government to ensure integrated and effective counterterrorism efforts, and coordinates and supports the development and implementation of all U.S. government policies and programs aimed at countering terrorism overseas.

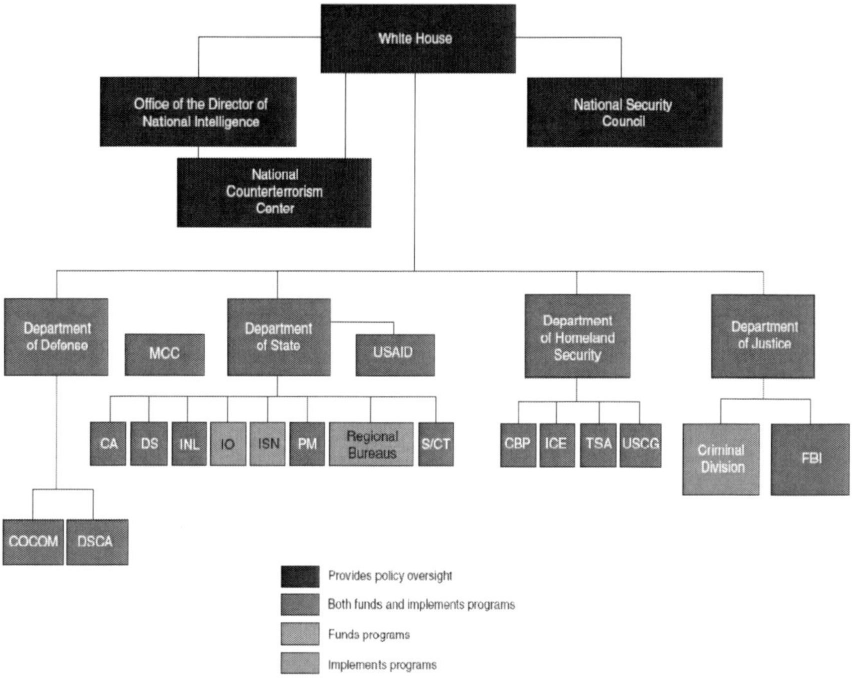

Source: GAO analysis of agency data and information.

Development; CA is Bureau of Consular Affairs; DS is Bureau of Diplomatic Security; INL is Bureau of International Narcotics and Law Enforcement Affairs; IO is Bureau of International Organization Affairs; ISN is Bureau of International Security and Nonproliferation; PM is Bureau of Political- Military Affairs; S/CT is Office of the Coordinator for Counterterrorism; CBP is U.S. Customs and Border Protection; ICE is U.S. Immigration and Customs Enforcement; TSA is Transportation Security Administration; USCG is United States Coast Guard; FBI is Federal Bureau of Investigation; COCOM is Combatant Command; and DSCA is Defense Security Cooperation Agency. In addition, the Human Smuggling and Trafficking Center serves as the focal point for interagency efforts to integrate and disseminate intelligence and information related to terrorist travel.

Figure 1. Key Agencies Providing Policy Oversight and Funding and/or Implementing Capacity-Building Programs Related to Preventing Terrorist Travel Overseas.

U.S. Government Assessments Have Identified Four Key Gaps in Foreign Countries' Capacity to Prevent Terrorist Travel Overseas

As shown in table 1, the U.S. government has identified four key gaps in foreign countries' capacity to prevent terrorist travel overseas.[6]

Table 1. Illustrations of Key Gaps in Foreign Partners' Capacity to Prevent Terrorist Travel Overseas

Key gaps	Illustrative examples
Sharing information about known and suspected terrorists	• Lack of a database system with terrorist screening information • Lack of information sharing between countries, including of terrorist screening information • Lack of access to databases with biometric information onknown or suspected terrorists
Addressing the use of fraudulent travel documents	• Manufacture and use of fraudulent travel documents • Lack of universal reporting to INTERPOL of lost or stolen passports, as well as some governments' limited use of this information
Ensuring passport issuance security	• Easily counterfeited or doctored low-quality passports from some countries • Continued validity of less secure passports for up to 10 years from issuance
Combating corruption in passport issuance and immigration agencies	• Corruption in passport issuance agencies facilitates fraudulent use of official travel documents and access to blank passports that could be used to make fraudulent passports • Corruption in immigration agencies allows terrorists to pass through checkpoints without being checked or arrested

Source: GAO analysis of NCTC and HSTC documents.

Note: Terrorist screening information includes identifying or biographical information—such as name, date of birth, and passport number—on known and suspected terrorists. INTERPOL—International Criminal Police Organization—facilitates cross-border police co-operation, and supports and assists all organizations, authorities, and services whose mission is to prevent or combat international crime.

HSTC and NCTC vulnerability assessments have identified sharing information about known and suspected terrorists as one key gap in foreign partners' capacity to prevent terrorist travel. For example, some countries do not have their own database systems with terrorist screening information or access to other countries' terrorist screening information, which contains biographical or biometric information about individuals who are known or suspected terrorists. Even when countries have terrorist screening information, they may not have reciprocal relationships to share such information or other travel-related information, such as airline passenger lists, with other countries, thereby limiting their ability to identify and prevent the travel of known and suspected terrorists. In addition, some countries do not have access to or fully use biometric information, which provides a unique identifier for each person, such as a fingerprint. For example, Pakistan has a centralized fingerprint database, but it is not shared across all law enforcement agencies, making the database less comprehensive and, as a result, more difficult for Pakistani government officials to prevent potential terrorists from traveling.

A second key gap in foreign partners' capacity relates to their ability to address the use of fraudulent travel documents. For instance, in many countries, fraudulent travel documents, including fraudulent passports and visas, are easy to obtain, and could thereby be used by people who want to travel under a false identity. In addition, some countries' failure to consistently report lost or stolen passports to the International Criminal Police Organization (INTERPOL) or to access INTERPOL's database that stores information on lost and stolen passports, can facilitate the use of legitimate passports by imposters. According to U.S. embassy officials in Kenya we spoke with, this is a common occurrence in Kenya, where individuals with a similar appearance to a Somali-American with a legitimate travel document will fraudulently use this travel document for illicit travel. Another common issue related to fraudulent travel documents is using fraudulent "breeder documents," such as birth certificates and drivers' licenses issued to support a person's false identity, to obtain genuine passports. The issue of fraudulent documents is further compounded by the lack of a requirement for a visa to some countries. For example, according to a former Pakistani official that had responsibilities related to immigration enforcement, fraudulent British passports are the most prevalent type of fraudulent travel document in Pakistan. Since British citizens are not required to obtain visas to travel to many countries, a terrorist could use one of these fraudulent passports to travel to many countries without further background checks that would occur through a visa adjudication process.[7]

The third key gap identified in the NCTC and HSTC assessments is a lack in some countries' abilities to ensure the security of their passport issuance systems. The passports from some countries are of low quality and are therefore easily stolen or counterfeited. For example, 18 countries still use passports that are not machine readable and almost half of countries use passports without biometric information stored electronically inside the passport. Such biometric information can include facial and fingerprint data, and can be used to authenticate the identity of travelers. In addition, once countries convert to biometric passports, previously issued passports may be valid for up to 10 years from their issuance dates.[8]

A fourth key gap in some foreign countries' capacity to prevent terrorist travel is in combating corruption in passport issuance and immigration agencies. Corruption in government agencies relevant to travel can facilitate the illicit travel of terrorists or other criminals. For example, corruption in passport issuance agencies can allow potential terrorists to obtain genuine passports under a false identity or blank passports that can be easily manipulated. U.S. embassy officials in Kenya told us that such false passports can be obtained for just a few hundred dollars in some cases. Further, corruption within countries' immigration agencies, such as border patrol or civil aviation officials with immigration duties, leaves a country's immigration system vulnerable to human smugglers and traffickers who often have established relationships with these corrupt officials. For example, according to U.S. embassy officials in Kenya, illicit travel facilitators are known to stand outside the airport and indicate to corrupt immigration officials through the window which individuals they should let pass without checking their passports. In addition, according to the HSTC terrorist travel vulnerability assessments, countries that are known for having corrupt immigration officials are more likely to be used by terrorists as transit countries so that the terrorists can avoid interdiction.

U.S. AGENCIES CONDUCT FOREIGN CAPACITY-BUILDING EFFORTS RELATED TO THREE OF THE FOUR KEY GAPS, BUT COORDINATION COULD BE IMPROVED

U.S. government foreign capacity-building programs and activities address to some degree most of the key gaps identified by the U.S. government in foreign governments' ability to prevent terrorist travel overseas.

Table 2. U.S. Government Programs and Activities That Address the Key Gaps in Foreign Countries' Capacity to Prevent Terrorist Travel Overseas

Key gap	Funding U.S. government agencies/ bureaus	Related programs and activities
Sharing information about known and suspected terrorists	State/S/CT	• Terrorist Interdiction Program – provides foreign immigration officials with a system to screen for potential terrorist travelers
	State/INL	• ICITAP– funds DOJ to provide unique database systems to countries tohelp them screen for potential terrorist or criminal travelers
	State/INL and State/ISN	• International Visitors Program – funds CBP to arrange briefings and U.S. visits by foreign officials to learn about U.S. management of terrorist screening information
	DOJ/FBI and State/S/CT	• Negotiating and maintaining agreements with foreign countries to shareterrorist screening information
	DHS, with State/S/CT and the State Bureau of European and Eurasian Affairs	• Negotiations to share Passenger Name Records data to prescreen airline passengers against terrorist screening information
Addressing the use of fraudulent travel documents	State/S/CT and State/DS	• Anti-Terrorism Assistance program – provides training in fraudulent travel document recognition
	State/INL	• Provides funding for training in fraudulent travel document recognition • International Visitors Program – funds CBP to arrange briefings and U.S. visits by foreign officials to learn about fraudulent travel document recognition
	State/DS	• Provides training in fraudulent travel document recognition
	DHS/ICE	• Provides training in fraudulent travel document recognition
	DHS/TSA	• Provides training in fraudulent travel document recognition
	DHS/CBP	• Provides training in fraudulent travel document recognition
	State/Consular Affairs and DHS/Office of Policy	• Encourages foreign countries to report lost or stolen passports to INTERPOL and works to enhance international standards and procedures for this reporting

Key gap	Funding U.S. government agencies/ bureaus	Related programs and activities
Ensuring passport issuance security	State/Consular Affairs	• Undertakes diplomatic efforts through the International Civil AviationOrganization to encourage other countries to use machine-readable and biometric passports • Provides training to delegations from foreign countries on passport issuance security • With State/INL, plans to provide passport anti-fraud training
	USAID	• Provided technical assistance to Paraguay to reform its passport issuance system
Combating corruption in passport issuance and immigration agencies	—	• No efforts are directly aimed at these agencies

Source: GAO analysis of agency data.

Note: State is Department of State; S/CT is Office of the Coordinator for Counterterrorism; INL is Bureau of International Narcotics and Law Enforcement Affairs; ICITAP is International Criminal Investigative Training Assistance Program; DOJ is Department of Justice; ISN is Bureau of International Security and Nonproliferation; CBP is U.S. Customs and Border Protection; FBI is Federal Bureau of Investigation; DHS is Department of Homeland Security; DS is Bureau of Diplomatic Security; ICE is U.S. Immigration and Customs Enforcement; TSA is Transportation Security Administration; USAID is U.S. Agency for International Development.

As shown in table 2, three of the four key gaps—sharing information about known and suspected terrorists, addressing the use of fraudulent travel documents, and ensuring passport issuance security—have been the subject of some programs and activities.

However, with regard to U.S. programs addressing the use of fraudulent travel documents, we found potential for overlap and duplication of effort, as multiple agencies that fund and implement numerous training courses do not always coordinate their activities.

While the U.S. government has many efforts aimed at helping foreign countries to combat corruption, none focus on the fourth gap of corruption related to passport issuance and immigration agencies.

U.S. Foreign Capacity-Building Efforts Address Sharing Information about Known and Suspected Terrorists

Multiple federal efforts are aimed at improving information sharing about known and suspected terrorists. First, State/S/CT's Terrorist Interdiction Program (TIP) enables immigration officials in countries at risk of terrorist activity to identify the attempted travel of known or suspected terrorists through the provision of a computerized system called the Personal Identification, Secure Comparison, and Evaluation System (PISCES). TIP provides participating countries with the PISCES software, hardware, and equipment to operate the software; any required maintenance and expansion of the system; and training on how to use it. During fiscal year 2010, the PISCES system processed an estimated 150,000 travelers per day entering or exiting 17 participating countries through ports of entry with PISCES installations.[9] In fiscal year 2010, State began to upgrade the PISCES software with biometric capabilities that further enhance host countries' capacity to interdict terrorists attempting to travel under a false identity.

Second, State's Bureau of International Narcotics and Law Enforcement Affairs (INL) has funded at least two projects to provide different types of database systems to foreign law enforcement authorities to help them screen for potential terrorist or criminal travelers. These projects are implemented through the DOJ/Criminal Division's International Criminal Investigative Training Assistance Program (ICITAP), a broad law enforcement development program that caters its program offerings to fit the host country's needs. First, in Bosnia and Herzegovina, ICITAP has provided the State Police Information Network to Bosnian border officials to allow them to link to INTERPOL databases to identify criminals who could then be denied entry to the country. Second, ICITAP has provided a separate system, the Total Information Management System, to Albania to enhance the country's capacity to screen for known terrorists. According to State, the governments of Kosovo and Albania are discussing adapting certain elements of the Total Information Management System for use in Kosovo as well.

Third, INL and State's Bureau of International Security and Nonproliferation have provided funding to DHS' U.S. Customs and Border Protection (CBP) to arrange trips for foreign officials to come to the United States to learn about how CBP uses and analyzes terrorist screening information. These trips are organized through the International Visitors Program, through which CBP arranges briefings and visits to CBP operations in the United States by foreign high-level customs and other law enforcement officials who perform or manage functions similar to those encompassed within CBP's area of responsibility and expertise. In fiscal year 2010, CBP organized 22 visits by foreign officials for this purpose.

Fourth, the United States enhances other countries' ability to prevent terrorist travel abroad by sharing terrorist screening information with other countries. Under Homeland Security Presidential Directive 6 (HSPD6), the Terrorist Screening Center[10] within the DOJ's Federal Bureau of Investigation (FBI) and the Terrorism Information Sharing Office within State/S/CT negotiate agreements with foreign countries to systematically share terrorist screening information, thereby enhancing both countries' abilities to prevent terrorist travel abroad through immediate and systematic access to information on known and suspected terrorists. Once the United States has signed an HSPD-6 agreement with a foreign country, the Terrorist Screening Center then shares the information agreed to with the foreign partners. As of May 2011, the Terrorist Screening Center shared terrorist screening information with 23 foreign countries.[11] In addition to the systematic information sharing on known and suspected terrorists that occurs through HSPD-6 agreements, the Terrorist Screening Center also has had approximately six one-time arrangements for sharing terrorist screening information with countries hosting special events.

Fifth, DHS leads an interagency negotiating team, on which State/S/CT and State's Bureau of European and Eurasian Affairs also serve, that is involved in renegotiating a 2007 agreement between the United States and the European Union on the exchange of Passenger Name Records data. Once a country has the capacity to analyze this type of information provided by airlines on its passengers, the country is able to prescreen airline passengers against terrorist screening information, thereby helping them to prevent terrorists from traveling abroad. The European Union is now considering developing such a system and CBP has hosted officials from the European Union for briefings on how the United States analyzes Passenger Name Records data.

According to State and DOJ officials, capacity-building efforts related to information sharing about known and suspected terrorists face some challenges. Some countries have expressed concerns about the privacy and protections related to the sharing of sensitive terrorist screening information.[12] For example, European countries that have negotiated HSPD-6 agreements with the United States have been concerned about data protection, redress, and privacy policies and procedures in both utilizing terrorist screening information from the United States and sharing terrorist screening information with the United States because of differences between U.S. and European laws. According to officials from the Terrorist Screening Center, such differences can include the countries' statutes of limitations that delineate how long they can keep derogatory information. According to State officials, another related challenge is that providing information to foreign countries involves a loss of control over the information and creates the possibility that the information could be compromised through internal corruption. To address both challenges, the United States and the foreign governments negotiate on specific information-sharing mechanisms and protections that are feasible and acceptable to both sides.

Several U.S. Foreign Capacity-Building Efforts Address the Use of Fraudulent Travel Documents but Some Efforts Lack Coordination

Multiple Agencies Fund and Implement Training Courses in Fraudulent Travel Document Recognition

Seven different U.S. government entities across three federal agencies are involved in providing fraudulent travel document training to foreign government officials, as shown in figure 2. In delivering the training, agencies have similar objectives and often provide the training to the same populations (e.g., immigration officials and law enforcement officials) to develop their skills in recognizing the characteristics of altered, counterfeit, or other fraudulent travel documents.

U.S. law enforcement officials working overseas from DHS/ICE and State's Bureau of Diplomatic Security (DS) provide the bulk of training in the recognition of fraudulent travel documents to foreign immigration and law enforcement officials. Specifically, attachés from DHS/ICE and in-country representatives from State/DS provide such training under the dual objectives of preventing terrorist travel and protecting U.S. interests. For example, in fiscal year 2010, ICE attachés provided 360 training courses, briefings, and outreach sessions on fraudulent travel document recognition and State/DS staff posted overseas provided 458 related training courses.[13]

In addition, State/S/CT and State/DS implement the Anti-Terrorism Assistance (ATA) program, which focuses on building foreign law enforcement officers' counterterrorism capabilities. ATA provides fraudulent travel document recognition training as part of achieving program goals related to preventing terrorist travel abroad.[14] In fiscal year 2010, 12 of the more than 350 courses provided by ATA were fraudulent travel document recognition courses. These courses were provided to law enforcement officials from 17 of the approximately 60 countries that received ATA training in fiscal year 2010.

Sources: GAO analysis of agency data and information; Corel and Art Explosion (clip art).

Figure 2. U.S. Agencies and Bureaus Involved in Providing Fraudulent Travel Document Recognition Training to Foreign Immigration and Law Enforcement Officials.

Other U.S. foreign capacity-building programs have implemented fraudulent travel document recognition courses, although their missions are not directly related to preventing terrorist travel abroad.

- State/INL provides funding for U.S. law enforcement agencies, including ICE, CBP, and the FBI, to implement the International Law Enforcement Academies (ILEA), which provide a general law enforcement training program that also includes some specialized training on how to combat certain criminal activities, including fraudulent travel documents. In fiscal year 2010, the ILEAs provided two courses specifically on fraudulent travel document recognition to law enforcement officials from 13 countries, as well as having training on this topic provided by ICE as part of the general law enforcement training offered at the ILEA in San Salvador that was delivered five times that fiscal year.[15] In addition, State/INL has provided funding to multiple entities to provide training in fraudulent travel document recognition. First, State/INL provides funding to CBP for related training, such as for fraudulent travel document training provided to Moroccan officials in fiscal year 2010 and for CBP's International Visitors Program, which, in fiscal year 2010, arranged six trips to the United States for foreign officials to learn how to recognize fraudulent travel documents. Also, State/INL has provided funding to the Organization of American States to deliver training in fraudulent document recognition throughout the Western Hemisphere and to the United Nations Office on Drugs and Crime to develop a manual on how to examine travel documents to determine their authenticity.

- The Transportation Security Administration (TSA) within DHS funds Aviation Security Sustainable International Standards Teams, which build select countries' aviation security through related training, technical assistance, and overall security assessments, in cases when these countries are having difficulty meeting International Civil Aviation Organization (ICAO) aviation security standards. In fiscal year 2010, as part of this effort, TSA funded one fraudulent travel document training course in Liberia, which was taught by ICE and CBP, as part of fulfilling that country's needs to meet ICAO standards related to detecting fraudulent travel documents.[16]
- CBP's Office of International Affairs has funded some fraudulent travel document recognition training related to its mission to enhance international border security. In fiscal year 2010, CBP funded one course in fraudulent document recognition for Mexican law enforcement officials.[17]
- In addition to training provided by ICE attachés, ICE's Office of International Affairs funds some additional fraudulent travel document recognition training courses, which involve ICE officials traveling from Washington, D.C., to instruct the courses. In fiscal year 2010, ICE funded four such training sessions for representatives from at least nine countries.

- Finally, the FBI has at times been involved in the provision of fraudulent travel document recognition training to foreign law enforcement officials, although it did not fund or implement any such training in fiscal year 2010. In March 2011, the FBI organized a training session for Indonesian officials in that country's police, state intelligence, public corruption commission, customs, immigration, military, and prosecutor's offices, a portion of which involved fraudulent travel document training that was provided by ICE and State/DS.

Agencies Funding and Implementing Fraudulent Travel Document Training Sometimes Lack Coordination

Our past work on issues that cut across multiple agencies shows that without a coordinated approach, programs can waste scarce funds and limit the overall effectiveness of the U.S. government's efforts.[18] GAO has found that, while collaboration among federal agencies can take different forms, practices that generally enhance collaboration include agreeing upon agency roles and responsibilities and identifying and addressing needs by leveraging resources.[19] GAO has further suggested that program officials require sufficiently detailed information to enable them to carry out their duties and responsibilities effectively, while collaborating when necessary to increase their efficiency.[20]

State/S/CT officials told us they were unaware of how many agencies and subagencies are involved in providing fraudulent travel document training to foreign officials, and they had not developed any mechanism to encourage coordination among all the parties involved. At the country level, we found that agency officials at two of the posts we visited did not always collaborate on the delivery of fraudulent travel document recognition training.[21] As a result, some planned training was duplicative and did not make an effective use of limited resources. For example, during our March 2011 visit to Pakistan, we identified two agencies planning to provide fraudulent travel document recognition training courses in April 2011 to Pakistani officials from the same agency without coordinating with one another. The ICE attaché planned one course that had a full roster of students but lacked funding, while ATA was simultaneously planning to hold two fully-funded fraudulent travel document courses in the same month although they had no students signed up for either course. Meanwhile, the ICE attaché had been certified through a train-the-trainer course provided by ICE's Forensic Document Laboratory to be an instructor for fraudulent travel document recognition courses. Since

ATA program officials were unaware of the existence of this local resource, the ATA program was still attempting to find two instructors from ICE to travel to Pakistan to teach the courses they were planning. In addition to potentially adding to program costs by not using the locally available instructor, this lack of coordination also could have unnecessarily increased demand on the Forensic Document Laboratory's resources. The Forensic Document Laboratory is one of the primary sources of instructors for ATA courses in fraudulent travel documents. Officials from the Forensic Document Laboratory in Washington, D.C., told us they provide train-the-trainer courses to make up for their lack of sufficient staff to fulfill all the training requests from overseas programs like ATA.

In Kenya, we found that representatives from two U.S. agencies, State and DHS, deliver fraudulent travel document training but do not collaborate. The ATA program, which is run by a contractor hired by State/DS in Kenya, provided approximately one course per year from fiscal year 2007 to 2010 in fraudulent travel documents to police and security officers, customs and immigration officers, forensic specialists, and training officers. A representative of State/DS posted overseas also provides many training courses in fraudulent travel documents for immigration officials. The CBP attaché, who represents DHS at the post, has provided many training courses on this topic to airport and border officials, as well as speaking on the topics of fraudulent travel documents, imposter recognition, and human trafficking to students in the Kenyan Immigration Service's basic training. Despite these three representatives providing this similar training, a representative from one of the agencies stated that although he coordinated with other countries providing similar training in Kenya, he did not do so with other U.S. agencies.

Agencies Also Address the Use of Fraudulent Travel Documents through Capacity-Building Efforts Related to Lost and Stolen Passports

State's Bureau of Consular Affairs attempts to build foreign partners' capacity to address the issue of fraudulent travel documents by encouraging countries to report lost and stolen passports to INTERPOL and to access INTERPOL's database to check against travelers arriving at ports of entry to identify and interdict people misusing passports.[22] According to INTERPOL, as of June 2011, the total number of countries contributing lost and stolen passport information was 158; and some of these have connected border checkpoints to INTERPOL's system for automated checking against its database. To facilitate the interdiction of people misusing lost and stolen passports, Consular Affairs also assisted in the drafting of a set of global standards for national management of lost and stolen passport data, which was provided to ICAO for adoption as a part of the global travel document standards.

DHS' Office of Policy has also played a role in enhancing other countries' capacity to report information about lost and stolen passports. First, they have participated in ongoing efforts to revise INTERPOL's procedures for the reporting of lost and stolen passport information to enhance the capabilities and compliance of such reporting by INTERPOL members. Similarly, to improve foreign partners' ability to detect fraudulent travel documents, DHS' Office of Policy has provided technical assistance towards the development of a pilot program to enhance the sharing of information related to fraudulent document alert data between members of the Group of Eight[23] and INTERPOL.

State and USAID Undertake Efforts to Improve Foreign Governments' Passport Issuance Security

Two agencies, State and USAID, have undertaken foreign capacity-building activities to improve other countries' passport issuance security. State's Bureau of Consular Affairs, with its mission of issuing secure U.S. passports to traveling Americans, is involved in some efforts to enhance foreign countries' passport issuance security.

- Consular Affairs has contributed to diplomatic efforts through ICAO to promote other countries' use of machine-readable passports and passports with biometric features. For example, it was involved in the development and promotion of ICAO's standards for machine-readable passports published in September 2006. These standards are related to a requirement that countries use machine-readable passports by April 2010, and also provided specifications for biometric enhancements that could be made to electronic passports.
- Consular Affairs has also, since 2009, provided briefings to representatives from over 50 passport issuance authorities on the elements of secure passports. For example, in 2010, Consular Affairs organized the training of a delegation from Turkey's passport office in Washington, D.C., which included briefings and organized tours of the Washington Passport Agency and the U.S. Government Printing Office.

- State/INL is funding Consular Affairs to provide passport antifraud training to officials from foreign passport issuance agencies, which will first be piloted in fall 2011. This training is designed to improve the integrity of other countries' passports and passport issuance by helping them institute organizations, processes, and procedures for detecting fraudulent passport applications as part of their adjudication and issuance processes.[24]

In addition, USAID provided technical assistance to the Paraguayan Ministry of Interior and National Police to reform Paraguay's identification system, including its national identity cards and passports.[25] According to USAID, the prior identification system in Paraguay was not in compliance with international security standards and was vulnerable to corruption. Implementation of the new integrated national identity card and passport system involved providing information technology improvements, as well as training on how to collect citizens' biometric data and on how to manage the new system. Entries in the new national database now include biometric identifiers, including fingerprints, photographs, and signatures, all of which are automatically verified upon entry into the database for their compliance with international standards. In addition, passports were redesigned and upgraded to ICAO requirements, resulting in more secure documents that are less susceptible to fraud.

The U.S. Government Has Anticorruption Efforts Overseas although Not Specifically Aimed at Passport Issuance and Immigration Agencies

While the U.S. government, through USAID and Millennium Challenge Corporation (MCC) anticorruption foreign capacity-building programs and State-led diplomatic efforts, has many efforts aimed at helping foreign countries to combat corruption, no U.S. government effort focuses directly on combating corruption in countries' passport issuance and immigration agencies.

USAID has developed a wide range of programs for fighting corruption, often fit to the needs and opportunities of the recipient country.[26] Some USAID anticorruption programs focus on a few specific sectors, including tax collection, customs collection, and the financial sector. In addition, USAID also has programs that have a broader effect on combating corruption, such as civil society programs to increase public awareness, promote citizen involvement and participation, and encourage civil society oversight of government; programs to decentralize powers to local governments; rule of law programs to improve the justice sector and thereby the ability to prosecute corruption cases; and programs to build anticorruption agencies within foreign governments. While not specifically targeting passport and immigration agencies, these broad anticorruption programs may have a beneficial, indirect effect on these countries' abilities to combat corruption in passport issuance and immigration agencies, thereby indirectly helping to prevent terrorist travel abroad.

Similarly, MCC has multiple anticorruption efforts across the 38 countries to which the MCC provides assistance. These anticorruption efforts include encouraging countries to: pass stronger anticorruption laws, strengthen oversight institutions, open up the public policy-making process to greater scrutiny, and increase corruption-related investigations and prosecutions. Such efforts, although not directly focused on passport issuance and immigration agencies, also may have a beneficial, indirect effect on these countries' abilities to combat corruption in these agencies, thereby indirectly helping to prevent terrorist travel abroad.

In addition, State has been involved in diplomatic efforts to discourage corruption in foreign countries. Multilaterally, State has advocated for the implementation of the UN Convention against Corruption, which came into force in December 2005 and provides a comprehensive set of standards, measures, and rules that all countries can apply in order to strengthen their legal and regulatory regimes to fight corruption.

State has encouraged and provided financial support for the development and launch this year of a peer review process through which countries will show how they are complying with their commitments under the UN Convention. In many countries, as well as through regional workshops in Africa, State engages in efforts to encourage or support countries in combating corruption, such as through encouraging the investigation and prosecution of corruption cases.

While none of these efforts focus directly on passport issuance or immigration agencies, their goal is to strengthen overall the laws, institutions, and capacity to prevent and prosecute corruption, which, according to State, intend to impact the integrity and effectiveness of all government functions and agencies.

No Performance Measures to Gauge Governmentwide Progress in Closing Key Gaps in Foreign Partners' Ability to Prevent Terrorist Travel Have Been Established

Performance Measures Are Key Tools for Decision Makers but U.S. National Counterterrorism Strategies Lack Governmentwide Performance Measures Related to Foreign Capacity Building to Prevent Terrorist Travel

The U.S. government lacks performance measures to assess governmentwide progress in closing the key gaps in foreign partners' capacity to prevent terrorist travel overseas. Performance measurement enables decision makers to make informed policy and budget decisions.[27] At the national level, U.S. counterterrorism strategies lack performance measures related to capacity building to prevent terrorist travel. Similarly, neither State, DOD, DHS, DOJ nor USAID has established such measures to accompany their agencywide strategies. Components of some agencies have relevant performance measures at the program level, but they cover only one of the four key gaps. Without comprehensive measures that encompass all U.S. government agency efforts, the U.S. government cannot determine governmentwide progress in building foreign partners' capacity to prevent terrorist travel.

As we have previously reported, performance information is essential to enable decision makers to make informed decisions. Specifying performance metrics is one tool used in evaluating the effectiveness of government efforts.[28] Agencies can also use performance information to make various types of management decisions to improve programs and results. In addition, as we have also reported, many federal efforts transcend more than one agency.[29] Closing the gaps in foreign partners' capacity to prevent terrorist travel is an example of such an issue, since it involves efforts funded and implemented by several agencies. In such situations, we have reported that it is important to have full information on how cross-cutting goals will be achieved.

The Intelligence Reform and Terrorism Prevention Act of 2004 highlighted the importance of constraining terrorist travel and directed NCTC to submit a strategy that combined terrorist travel intelligence, operations, and law enforcement into a cohesive effort to intercept terrorists, find terrorist travel facilitators, and constrain terrorist mobility domestically and internationally. The resulting NCTC 2006 *National Strategy to Combat Terrorist Travel* lists some U.S. government activities related to helping partner nations build capacity to limit terrorist travel but contains no performance measures to assess governmentwide progress.

Similarly, the National Security Council, which coordinates national security and foreign policy among various U.S. government agencies, issued the *National Strategy for Combating Terrorism* in September 2006, which established the goal of disrupting terrorist travel internationally through various means, including building international capacity to secure travel and combat terrorist travel.[30] In June 2011, the President issued the *National Strategy for Counterterrorism*, which again highlighted the importance of enhancing the capacity of foreign partners to prevent terrorist travel across national borders. However, these unclassified strategies lack performance measures related to foreign capacity building to prevent terrorist travel.

Individual Agency Strategies Do Not Contain Performance Measures Related to Foreign Capacity Building to Prevent Terrorist Travel Abroad, but Some Agency Components Track Efforts Related to One Key Gap

We examined individual agency strategies for the agencies funding and/or implementing foreign capacity-building programs and activities related to preventing terrorist travel, including for State, DHS, DOD, DOJ, and USAID. We found that each agency's strategy acknowledged the important role the agency plays in combating international terrorism. However, none of the agencies' strategies contained performance indicators to measure progress related to helping countries close the key gaps in their ability to prevent terrorist travel.

Some agency components have made efforts to track the performance of their specific program efforts aimed at improving information sharing about known and suspected terrorists—one of the four key gaps. None of the agencies have performance measures related to the other three key gaps in foreign partners' capacity to prevent terrorist travel. Related to information sharing, State's S/CT and Director of U.S. Foreign Assistance have performance indicators for TIP that address sharing information on known and suspected terrorists.[31] In fiscal year 2009, S/CT created the performance indicator—the percentage of the highest priority countries capable of screening for terrorists through TIP/PISCES that receive biometric capabilities. The target for that performance indicator for fiscal year 2010 was that 50 percent of the 17 countries currently supported by TIP would have biometric capability. No fiscal year 2010 results have yet been publicly reported for this measure. The Director of U.S. Foreign Assistance's performance measure for TIP is the number of ports of entry supported by TIP. Figure 3 shows the increase in the number of ports of entry supported by TIP, and the annual targets, from 2006 to the present.

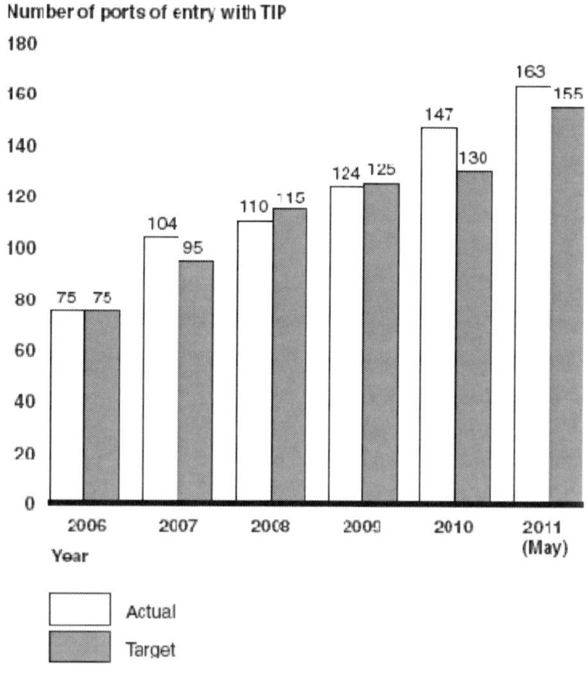

Source: GAO analysis of State Department data.

Figure 3. Number of Ports of Entry Supported by Terrorist Interdiction Program since 2006.

State's country-level plans also sometimes contain performance measures for U.S. counterterrorism efforts in that country. For example, State has performance measures in its 2012 mission strategic plans for Kenya and Thailand. For Kenya, the performance measure is—the government of Kenya should demonstrate capacity and resolve to prevent and respond to threats of terrorism by, among other things, expanding TIP/PISCES coverage to additional border crossings. For Thailand, the performance measure is—Thailand should develop effective export control and border security systems that meet international standards by installing new software for TIP/PISCES at targeted airport locations and expanding the program to new ports of entry.

Finally, DOJ/FBI also has two performance measures related to the information sharing gap that assess the Terrorist Screening Center's efforts to share terrorist screening information with foreign partners.[32] The FBI has not set targets for either of these measures.

Overall, these relatively narrow agency-specific measures that exist do not provide a comprehensive basis for assessing governmentwide progress in building foreign partners' capacity for two reasons. First, they necessarily focus on specific program efforts, not governmentwide progress. Second, they cover only one of the four key gaps in the capacity of foreign countries to prevent terrorist travel overseas.

CONCLUSION

Inhibiting the movement of terrorists across international borders is a key part of the U.S. strategy for protecting the United States and its interests abroad. Although agencies have implemented significant new domestic programs to prevent terrorists from entering the United States, events of the past few years illustrate that the international travel system is only as secure as its weakest link. As a result, the United States seeks to enhance the capacity of its foreign partners to prevent terrorist travel overseas, with agencies implementing a variety of programs and activities to close key gaps in our foreign partners' capacity. However, some of these efforts—such as improving foreign partners' capacity to prevent the use of fraudulent travel documents—are not always well coordinated and create the risk of duplication and overlap. In light of the limited resources available to address these important issues, it is critically important to ensure that such resources are used efficiently. Further, while more than 5 years have passed since the *National Strategy to Combat Terrorist Travel* linked our foreign partners' capacity to constrain terrorist travel to our own national security, the U.S. government still lacks an effective system for measuring and reporting progress toward the goal of enhancing our foreign partners' capacity. As agencies implement the new *National Strategy for Counterterrorism*, it is important to focus on measuring, tracking, and reporting on governmentwide progress toward the goal of enhancing foreign partners' capacity to prevent terrorist travel. Without such information, the U.S. government cannot efficiently assess the effectiveness of its efforts and planners and decision makers may lack information vital to addressing foreign policy needs and leveraging U.S. resources.

Recommendations for Executive Action

In order to institute a coordinated approach for delivering fraudulent travel document recognition training overseas to ensure that U.S. agencies prevent overlap and duplication; and given State's role in working with all appropriate elements of the U.S. government to ensure integrated and effective international counterterrorism efforts, we recommend that:

- State develop a mechanism for agencies involved in funding and implementing fraudulent travel document recognition training at overseas posts to coordinate the delivery of such training to foreign partners.

To allow the U.S. government to determine the extent to which it is building foreign partners' ability to prevent terrorist travel abroad and to make adjustments to improve its programs accordingly, we recommend that:

- The National Security Council, in collaboration with relevant agencies, develop a mechanism to measure, track, and report on U.S. progress across the government toward its goal of enhancing foreign partners' capacity to prevent terrorist travel.

Agency Comments and Our Evaluation

We provided a draft of this report to State, DHS, DOD, DOJ, the Department of Transportation, USAID, NCTC, and the National Security Staff of the National Security Council. DHS and State provided written comments. State, DHS, DOJ, and NCTC provided technical comments, which we incorporated where appropriate. DOD, the Department of Transportation, USAID, and the National Security Staff did not provide any comments on the draft.

In commenting on a draft of this report, State agreed with our recommendation that it should develop a mechanism to enhance coordination among the agencies involved in funding and implementing fraudulent travel document training overseas. State noted that efforts to enhance such coordination have begun at the country level, and that coordination in this area is also needed in terms of strategic, budget, and program planning at the agencywide and interagency levels. In addition, DHS, in its letter commenting on our report, indicated its commitment to working with other relevant agencies to stop terrorists from traveling across international borders, including through contributing to coordinated efforts to prevent any overlap and duplication.

Regarding our recommendation to the National Security Council to work with relevant agencies to develop a mechanism to measure, track, and report on governmentwide progress toward its goal of enhancing foreign partners' capacity to prevent terrorist travel, the National Security Staff did not provide any comment. However, in previous meetings with us, the National Security Staff acknowledged the need for such a mechanism.

Charles Michael Johnson Jr.
Director, International Affairs and Trade

APPENDIX I: SCOPE AND METHODOLOGY

In this report, we (1) identified the key gaps the U.S. government has assessed in foreign countries' capacity to prevent terrorist travel overseas, (2) evaluated how U.S. foreign capacity-building efforts address those gaps, and (3) assessed the extent to which the U.S. government is measuring progress in its efforts to close those gaps.

Our work focused on the efforts of the Departments of State (State), Homeland Security (DHS), Defense (DOD), and Justice (DOJ) to build foreign partners' capacity to prevent terrorist travel overseas. Within these agencies, we met with officials from several relevant components that are contributing to the U.S. government goal of enhancing foreign partners' ability to prevent terrorist travel, including: State's Office of the Coordinator for Counterterrorism (S/CT), Bureau of Diplomatic Security (DS), and Bureau of International Narcotics and Law Enforcement Affairs (INL); DHS's U.S. Immigration and Customs Enforcement, U.S. Customs and Border Protection, Transportation Security Administration (TSA), and Office of International

Affairs; and DOJ's Federal Bureau of Investigation and Criminal Division. We focused on these agencies and components as a result of our assessment of agency efforts noted in the *National Strategy to Combat Terrorist Travel*, our review of information in previous and ongoing GAO work in counterterrorism and aviation security, and discussions with U.S. agency officials regarding the agencies with which they collaborate.

To obtain examples of U.S. efforts and more in-depth understanding of specific participation in U.S. capacity-building programs designed to prevent terrorist travel overseas, we selected four countries in which to conduct field work. We selected Kenya, Pakistan, the Philippines, and Thailand, based on criteria that included: designation as a terrorist safe haven, presence of key U.S. agency personnel at post, and coverage of key regions to counterterrorism.[1] In each location, we met with U.S. government personnel involved in capacity building to prevent terrorist travel abroad to learn about the key gaps in those countries' abilities to prevent terrorist travel overseas, the types of capacity-building activities they undertake related to preventing terrorist travel, and how they measure progress and report results. We also met with foreign government officials in three of the four countries to learn about the challenges they face in improving their ability to prevent terrorist travel abroad and th perspectives on the effectiveness of U.S. efforts.

To identify what the U.S. government has assessed to be the key gaps in foreign partners' capacity to prevent terrorist travel overseas, we reviewed the NCTC and Human Smuggling and Trafficking Center's (HSTC) terrorist travel vulnerability assessments from 2005, 2008, and 2009. Based on interviews with the HSTC, we learned that these are the only comprehensive U.S. government assessments of vulnerabilities within the foreign travel system. We reviewed all three documents to identify the key gaps because, according to HSTC officials, each assessment is not comprehensive. Rather, they are additive, so the assessments taken together represent a full picture of the vulnerabilities. We performed our review of these assessments by noting instances when certain gaps, threats, vulnerabilities, or areas for improvement to the international travel system generally or related to specific foreign countries were discussed. For the purposes of this review, we considered gaps to be threats, vulnerabilities, and areas for improvement mentioned in the assessments. The parts of the assessments that identify vulnerabilities limited to the U.S. travel system were not included within our analysis since they did not relate to the scope of our review. To distinguish between the key gaps identified in these reports and other vulnerabilities identified that were not key gaps, we reviewed the frequency with which each gap/vulnerability was mentioned in the reports. The HSTC confirmed our summary of the key gaps and other vulnerabilities. We also consulted with agency officials at headquarters, the missions in our example countries, and the intelligence community to identify examples of the key gaps in each country and corroborate our findings.

To evaluate how U.S. foreign capacity-building programs address those gaps, we examined relevant documents including program descriptions, and agency- and program-level strategic documents, including the 2012 Mission Strategic and Resource Plans. We conducted interviews with agency officials from State, DHS, DOJ, DOD, the Department of Transportation, and the U.S. Agency for International Development (USAID), in Washington, D.C., and in our example countries where officials were involved in relevant capacity-building programs. We also interviewed officials from the NCTC and National Security Staff. To show the level of different agencies' involvement in the delivery of fraudulent travel document recognition training to foreign officials, we requested data from all relevant agencies on the number of such courses that they funded and implemented in fiscal year 2010. We determined that these data were sufficiently reliable for our purposes.

To assess the extent to which the U.S. government is measuring progress in its efforts to enhance foreign partners' ability to constrain terrorist travel overseas, we analyzed relevant U.S. planning and evaluation documents including the 2006 *National Strategy to Combat Terrorist Travel*, the 2006 *National Strategy for Combating Terrorism*, the 2008 *National Implementation Plan for the War on Terror*, and the 2011 *National Strategy for Counterterrorism*. We also reviewed the relevant agency strategic documents for State, DHS, DOD, DOJ, and USAID. The State documents included the fiscal year 2012 strategic and resource plans of the bureaus of S/CT, DS, INL and Consular Affairs as well as the fiscal year 2012 Mission Strategic and Resource Plans of our example countries. We determined that State's data on performance indicators for the Terrorist Interdiction Program were sufficiently reliable for our purposes.

To identify what have been the reported results of these efforts, we reviewed relevant agency reports including: State's *Annual Report on Assistance Related to International Terrorism* from fiscal year 2009, the strategic and resource plans of the bureaus of S/CT, DS, INL and Consular Affairs as well as the Mission Strategic and Resource Plans of our example countries, DOJ performance reports, and the DHS *Annual Performance Report* for fiscal years 2008–2010. We also discussed progress with agency officials at headquarters and at the missions of our example countries.

We conducted this performance audit from July 2010 to June 2011 in accordance with generally accepted government auditing standards. Those standards require that we plan and perform the audit to obtain sufficient, appropriate evidence to provide a reasonable basis for our findings and conclusions based on our audit objectives. We believe that the evidence obtained provides a reasonable basis for our findings and conclusions based on our audit objectives.

Appendix II: U.S. Capacity-Building Efforts Related to the Vulnerabilities of Aviation and Border Security

Multiple agencies are involved in many programs and activities to build the capacity of foreign countries to address vulnerabilities in their aviation and border security, as shown in table 3. Since countries can have both land and water borders, we include both land border and maritime security programs under border security. For both aviation and border security programs, we include only programs that include elements relating to preventing illicit passenger travel. We have not included other aviation or border security programs that focus only on preventing illicit cargo shipments.

Table 3. U.S. Government Programs and Activities That Address Other Vulnerabilities Identified in Foreign Countries' Capacity to Prevent Terrorist Travel Overseas

Areas of vulnerability	Funding U.S. government agencies/ bureaus	Related programs and activities
Aviation security	State/DS and State/S/CT	• Antiterrorism Assistance Program – provides training in airport security management and related quality control procedures
	DHS/TSA	• Aviation Security Sustainable International Standards Teams – provide teams of subject matter experts to conduct needs assessments and follow-up visits to deliver agreed-upon assistance, such as aviation security-related training and equipment provisions
	State/Bureau of African Affairs	• Provides funding to the Department of Transportation to administer the Safe Skies for Africa program, which provides funding to TSA to train airport security personnel in ICAO aviation security requirements and standards

Areas of vulnerability	Funding U.S. government agencies/ bureaus	Related programs and activities
	State/INL	• Provides some equipment for airport security, as well as some related training and technical assistance
	State/ISN	• Export Control and Border Security Program – funds a CBP-implemented course on observing airline passengers' behavior to look for irregularities
	State/INL and State/ISN	• International Visitors Program – funds CBP to arrange briefings and U.S. visits by foreign officials to learn about aviation security
	State/S/CT	• Regional Strategic Initiative – funds regional workshops in airport security technology, passenger anascreening procedures
	DOD/African Command	• Provided funding for airport interdiction training and donated airport screening equipment in Kenya
	State/Bureau of International Organization Affairs	• Contributes nearly $1 million each year to ICAO's technical security assistance programs
	State/Bureau of Western Hemisphere Affairs	• Provides funding to support the Organization for American States' aviation and airport security training for personnel throughout Latin America
	State/Bureau of East Asian and Pacific Affairs	• Provides funding to support Asia-Pacific Economic Cooperation projects related to aviation security, including a conference in Vietnam in 2009
Border security	DOD	• In collaboration with State/PM, DOD implements the Section 1206 Program, which trains and equips foreign military and nonmilitary maritime forces, including in border and maritime security
	State/DS and State/S/CT	• Antiterrorism Assistance Program – provides training in border control management and maritime port and harbor security management
	DHS/CBP	• Provides capacity building training and technical assistance in border contro practices
	State/INL	• Funds and implements maritime and border security programs that include training and equipment provisions
	State/INL and State/ISN	• International Visitors Program – funds CBP to arrange briefings and U.S. visits by foreign officials to learn about border and maritime security
	State/PM	Along with DOD/DSCA, State/PM implements the Foreign Military Financing program, which has provided equipment to foreign countries to strengthen their border security
	DOD/DSCA	Combating Terrorism Fellowship Program – provides targeted, nonlethal, combating terrorism education and training, which can include courses in maritime or border security

Table 3. (Continued)

Areas of vulnerability	Funding U.S. government agencies/ bureaus	Related programs and activities
	State/PM	Funds DOD/DSCA to implement the International Military Education and Training program, which provides professional military training to foreign militaries. This training can help improve border and maritime forces in countries where the military controls the border.
	DHS/U.S. Coast Guard	Builds the maritime capacities of foreign military and security services through training and technical assistance in areas such as law enforcement boarding and searching for stowaways.

Source: GAO analysis of agency data.

Note: State is Department of State; DS is Bureau of Diplomatic Security; S/CT is Office of the Coordinator for Counterterrorism; DHS is Department of Homeland Security; TSA is Transportation Security Administration; ICAO is International Civil Aviation Organization; INL is Bureau of International Narcotics and Law Enforcement Affairs; ISN is Bureau of International Security and Nonproliferation; CBP is U.S. Customs and Border Protection; DOD is Department of Defense; PM is Bureau of Political-Military Affairs; DSCA is Defense Security Cooperation Agency.

End Notes

[1] The mission of the NCTC is to lead the U.S. effort to combat terrorism at home and abroad by analyzing the threat, sharing that information with U.S. partners, and integrating all instruments of national power to ensure unity of effort.

[2] We originally planned fieldwork in Yemen in March 2011 but were unable to travel there due to the deteriorating security situation at the time.

[3] Pub. L. 108-458, 118 Stat. 3809, Sec. 7201(b).

[4] Pub. L. 110-53, 121 Stat. 346.

[5] As of January 2011, HSTC had 18 full-time and 2 part-time staff, mostly detailees from DHS.

[6] The NCTC and HSTC assessments also identified vulnerabilities in foreign countries' aviation security and border security, but to a lesser extent, so we do not include these vulnerabilities in the key gaps. However, we identify U.S. programs and activities that address vulnerabilities in aviation and border security in app. II.

[7] GAO, *Visa Waiver Program: DHS Has Implemented the Electronic System for Travel Authorization, but Further Steps Needed to Address Potential Program Risks*, GAO-11-335 (Washington, D.C.: May 5, 2011).

[8] GAO, *Border Security: Security of New Passports and Visas Enhanced, but More Needs to Be Done to Prevent Their Fraudulent Use*, GAO-07-1006 (Washington, D.C.: July 31, 2007).

[9] The countries participating are: Afghanistan, Cambodia, Cote D'Ivoire, Djibouti, Ethiopia, Ghana, Iraq, Kenya, Kosovo, Macedonia, Malta, Pakistan, Tanzania, Thailand, Uganda, Yemen, and Zambia.

[10] Pursuant to HSPD-6, the Terrorist Screening Center was established to consolidate the U.S. government's approach to terrorist screening and to provide for the appropriate and lawful use of terrorist information in screening processes.

[11] For more information on HSPD-6 agreements, see GAO-11-335.

[12] GAO-11-335.

[13] In addition to training foreign law enforcement and immigration officials, some of these training sessions provided by State/DS officers overseas also were provided to other groups, such as travel agencies, and included related training on analysis and recognition of trends and practices used by document vendors and human smugglers, such as schemes used to obtain visas for large groups of applicants.

[14] ATA is co-managed by State/S/CT and State/DS, with S/CT providing policy guidance and DS managing program operations. For more information about ATA, see GAO, *Combating Terrorism: State Department's Antiterrorism Program Needs Improved Guidance and More Systematic Assessments of Outcomes*, GAO-08-336 (Washington, D.C.: Feb. 29, 2008).

[15] There are ILEAs in San Salvador, El Salvador; Gaborone, Botswana; Bangkok, Thailand; Budapest, Hungary; and Roswell, New Mexico; all of which, according to INL officials, offer unique training programs catered to the needs of their regions.

[16] ICAO is a United Nations organization that sets standards and recommended practices for the safety, security, environmental protection, and sustainable development of air transport. Included in ICAO's guidelines for aviation security are provisions instructing countries to guard against the misuse of their travel documents and to facilitate detection of cases where such documents have been unlawfully altered, replicated, or issued.

[17] CBP, through its Carrier Liaison Program, also provides fraudulent travel document recognition training to civilian airport employees at foreign airports. In fiscal year 2010, CBP provided over 150 such training sessions.

[18] GAO, *Results-Oriented Government: Practices That Can Help Enhance and Sustain Collaboration among Federal Agencies*, GAO-06-15 (Washington, D.C.: Oct. 21, 2005). Also, see GAO, *Opportunities to Reduce Potential Duplication in Government Programs, Save Tax Dollars, and Enhance Revenue*, GAO-11-318SP (Washington, D.C.: Mar. 1, 2011).

[19] GAO-06-15.

[20] GAO, *Standards for Internal Control in the Federal Government*, GAO/AIMD-00-21.3.1 (Washington, D.C.: November 1999).

[21] In some embassies, a Law Enforcement Working Group meets to coordinate law enforcement activities, which can include the coordination of training to foreign officials. Law Enforcement Working Groups had been meeting regularly in Pakistan and Kenya at the time of our visits to these countries in March 2011 when we learned of these instances of coordination lacking in the delivery of fraudulent travel document recognition training.

[22] The Implementing Recommendations of the 9/11 Commission Act of 2007 requires that countries participating in the Visa Waiver Program enter into an agreement with the United States to report, or make available to the United States through INTERPOL or other means as designated by the Secretary of Homeland Security information about lost or stolen passports. As of January 2011, all 36 Visa Waiver Program countries were sharing lost and stolen passport information with the United States. DHS regularly monitors Visa Waiver

Program countries' compliance with this requirement. For more information, see GAO-11-335.

[23] The Group of Eight (G8) refers to the group of eight highly industrialized nations— France, Germany, Italy, Great Britain, Japan, United States, Canada, and Russia—which hold a yearly meeting, the G8 Summit. Meetings are intended to foster consensus on global issues such as economic growth and crisis management, global security, energy, and terrorism.

[24] According to State's Bureau of Consular Affairs, this passport antifraud training will also cover topics such as how to prevent malfeasance and account for vulnerable items such as passport issuance systems and blank passport books. As a result, this future training may in part also help to address the key gap in combating corruption in passport issuance and immigration agencies discussed in the next section.

[25] This assistance was provided as a part of the Millennium Challenge Corporation's Paraguay Threshold Program Stage I (2007-2009). For more information on the Millennium Challenge Corporation, see GAO, *Millennium Challenge Corporation: MCC Has Addressed a Number of Implementation Challenges, but Needs to Improve Financial Controls and Infrastructure Planning,* GAO-10-52 (Washington, D.C.: Nov. 6, 2009).

[26] See GAO, *Foreign Assistance: U.S. Anticorruption Programs in Sub-Saharan Africa Will Require Time and Commitment,* GAO-04-506 (Washington D.C.: Apr. 26, 2004).

[27] GAO, *Executive Guide: Effectively Implementing the Government Performance and Results Act*, GAO/GGD-96-118 (Washington, DC: June 1996).

[28] GAO/GGD-96-118.

[29] GAO, *Government Performance: GPRA Modernization Act Provides Opportunities to Help Address Fiscal, Performance, and Management Challenges*, GAO-11-466T (Washington, DC: Mar. 16, 2011).

[30] The National Security Council is the principal forum used by the President of the United States for considering national security and foreign policy matters with his senior national security advisors and cabinet officials and is part of the Executive Office of the President of the United States. The function of the Council is to advise and assist the president on national security and foreign policies. The Council also serves as the president's principal arm for coordinating these policies among various government agencies.

[31] Performance measures related to the key gap of information sharing about known and suspected terrorists are found in State/S/CT's bureau strategic plan, and in the Director of U.S. Foreign Assistance's master list of standard indicators.

[32] The FBI considers the content of these performance measures to contain sensitive information and therefore the exact measures cannot be discussed in this public report.

End Notes for Appendix I

[1] We originally planned also to conduct fieldwork in Yemen in March 2011, but were unable to travel there due to the deteriorating security situation at the time.

INDEX

D

E

F

G

H

I

J

K

L

M

N

O

P

Q

R

U

V

W

Y